SpringerBriefs on Case Studies of Sustainable Development

Series Editors

Asit K. Biswas, Third World Centre for Water Management, Los Clubes, Atizapan, Mexico
Cecilia Tortajada, Los Clubes, Atizapan, Mexico

The importance of sustainable development has been realized for at least 60 years, even though the vast majority of people erroneously think this concept originated with the Brundtland Commission report of 1987 on Our Common Future. In spite of at least six decades of existence, we only have some idea as to what is NOT sustainable development rather than what is. SpringerBriefs on Case Studies of Sustainable Development identify outstanding cases of truly successful sustainable development from different parts of the world and analyze enabling environments in depth to understand why they became so successful. The case studies will come from the works of public sector, private sector and/or civil society. These analyses could be used in other parts of the world with appropriate modifications to account for different prevailing conditions, as well as text books in universities for graduate courses on this topic. The series of short monographs focuses on case studies of sustainable development bridging between environmental responsibility, social cohesion, and economic efficiency. Featuring compact volumes of 50 to 125 pages (approx. 20,000–70,000 words), the series covers a wide range of content—from professional to academic—related to sustainable development. Members of the Editorial Advisory Board: Mark Kramer, Founder and Managing Director, FSG, Boston, MA, USA Bernard Yeung, Dean, NUS Business School, Singapore

More information about this series at http://www.springer.com/series/11889

David B. Brooks · Julie Trottier ·
Giulia Giordano

Transboundary Water Issues in Israel, Palestine, and the Jordan River Basin

An Overview

 Springer

David B. Brooks
University of Victoria
Ottawa, ON, Canada

Giulia Giordano
EcoPeace Middle East
Tel Aviv, Israel

Julie Trottier
ART-Dev, Site Saint-Charle
Directrice de Recherche CNRS
Montpellier, France

ISSN 2196-7830   ISSN 2196-7849   (electronic)
SpringerBriefs on Case Studies of Sustainable Development
ISBN 978-981-15-0251-4   ISBN 978-981-15-0252-1   (eBook)
https://doi.org/10.1007/978-981-15-0252-1

This Springer imprint is published by the registered company Springer Nature Singapore Pte Ltd.
The registered company address is: 152 Beach Road, #21-01/04 Gateway East, Singapore 189721, Singapore

Science may indeed provide us with true opinion concerning certain aspects of human nature and the natural world so that we can choose a rule of life that does not flout reality. But it cannot tell us what reality ultimately is, and it cannot choose the rule for us.
Ophuls (2011, p. 134–135)[1]

[1]Ophuls, W (2011) *Plato's Revenge: Politics in an Age of Ecology*. The MIT Press, Cambridge, MA, USA.

This book is dedicated to all those women and men who work to create programs and policies that demonstrate that peace for the nations of Palestine, Jordan, and Israel is possible, and also those who live in the region and ignore or oppose people who say that the other side does not want peace.

Acknowledgements

Some portions of this book were originally published as A New Paradigm for Transboundary Water Agreements: The Opportunity for Israel and Palestine, Chap. 6 in *Management of Transboundary Water Resources under Scarcity: A Multidisciplinary Approach*, edited by Ariel Dinar and Yacov Tsur. London: World Scientific, pp.159–187. Other portions, first appeared in an article, also co-authored with Julie Trottier, entitled De-nationalization and De-Securitization of Transboundary Water Resources and published in the *International Journal of Water Resources Development*, vol. 30 (2), pp. 211–223. The ideas, though not necessarily the words, in Chap. 4, first appeared in other publications of EcoPeace Middle East. We also thank the three offices of EcoPeace Middle East—Amman, Jerusalem, Tel Aviv–for statistical, graphical, and editorial work in preparing this and previous publications. We must also thank the faculty and the participants of The John Harvard Seminar, *Topographies of Citizenship*, at Cambridge University in the UK, at which early versions of this book, especially Chap. 4, were reviewed and critiqued. We also thank Diane Beckett for help with organization and references. Finally, David must also thank Conversa Language Services in Ottawa, Canada, where David has an office, for providing a most agreeable place for him to work.

Contents

About the Authors

Dr. David B. Brooks was educated in geology and economics and spent much of his professional career with Canada's International Development Research Centre. His main research interests are split between water soft paths in North America and water demand management in the Middle East. Among his books are *Watershed: The Role of Fresh Water in the Israeli-Palestinian Conflict* (IDRC Books, 1994, co-author); and *Making the Most of the Water We Have: The Soft Path Approach to Water Management* (Earthscan, 2009, co-editor). In 2012, Dr. Brooks received an honorary doctorate of environmental studies from the University of Waterloo.

Dr. Julie Trottier is a director of research at France's National Center for Scientific Research (CNRS). With formal studies in chemistry, politics, and Islamic studies, she has focused her research for the last 20 years on the politics of water in Israel and Palestine. She published *Water Politics in the West Bank and Gaza Strip* (PASSIA 1999) and *Water Management, Past and Future* (Oxford University Press 2004). She directs the research project *"Managing the Paracommons of Palestinian Water"*.

Dr. Giulia Giordano is International Affairs Manager at EcoPeace Middle East, a trilateral organization based in Israel, Jordan, and Palestine. Formerly, she was a visiting researcher at the Hebrew University of Jerusalem and a fellow lecturer at Al-Quds University in East Jerusalem, where she taught two courses in Human Rights and International Environmental Law. She received her Ph.D. in Cooperation for Peace and Development from the University "Stranieri" of Perugia, Italy.

Abbreviations and Acronyms

ATCA   Anti-Terrorism Clarification Act (USA)
BWC   Bilateral Water Commission
BWT   Boundary Waters Treaty (USA and Canada)
CA   Civil Administration (Israeli Management on the West Bank)
EROI   Energy Return on Investment
FAO   Food and Agriculture Organization of the United Nations
GDP   Gross domestic product
GHG   Greenhouse gases
IJC   International Joint Commission (USA and Canada)
INSS   Institute for National Security Studies (Israel)
IPCRI   Israel–Palestine Center for Research and Information
JVA   Jordan Valley Authority (Jordan)
JWC   Joint Water Committee (Israel and Palestine)
kWh   Kilowatt-hour
MCM   Million cubic metres
MENA   Middle East and North Africa
MoU   Memorandum of Understanding
MOWI   Ministry of Water and Irrigation (Jordan)
NGO   Non-Governmental Organization
OECD   Organisation for Economic Co-operation and Development
PASSIA   Palestinian Academic Society for the Study of International Affairs
PNA   Palestinian National Authority
PWA   Palestinian Water Authority
RSDSWC   Red Sea-Dead Sea Water Conveyance
TFDD   Transboundary Freshwater Dispute Database (Oregon State
          University)
TVA   Tennessee Valley Authority (USA)
USAID   US Agency for International Development
WAFA   Palestine News Agency; also Palestinian News and Information
          Agency

WEN   Water-Energy Nexus
WMB   Water Mediation Board
WRWG   Water Resources Working Group (Multilateral Track of the Middle
               East Peace process)
WSRC   Water Sector Regulatory Council (PWA)
WWTP   Wastewater treatment plant

List of Figures

List of Boxes

Chapter 1
Global and Regional Perspectives

Abstract This book builds upon the EcoPeace Proposal for a water agreement between Israel and Palestine, together with those parts of the Proposal that involve the western portions of Jordan. Chapter One identifies key aspects of the global and regional setting for the review of current and possible future water management in Israel, Palestine, and Jordan. It indicates how such an agreement must reflect the role of water as a flow rather than as a stock, distinguish between water use and water consumption, and recognize the numerous benefits from transboundary water management among the three countries. Thanks to the Annex 2 of the Peace Treaty between Israel and Jordan, joint management of their shared water has been working well, though future relationships are uncertain at present. However, existing arrangements for Israel and Palestine are, at best, inadequate and, in some cases, counterproductive. This Chapter concludes with a description of the organization of the rest of the book.

1.1 Introduction

Although resolution of issues related to fresh water shared by Israel and Palestine will not alone bring about peace between the two peoples, in the absence of a just resolution of water issues, no peace can be complete. Further, in the absence of *sustainable* use of water by both peoples and those living or working in the Jordan River basin, social and economic development will be threatened, and so too will peace for the region.

Though obviously essential to any final status agreement between Israel and Palestine, remarkably little attention has been devoted to the design of the water components of the agreement, and even less to their implementation. Clearly, agreements must be flexible and dynamic rather than rigid and static. Annex 2 of the Israel-Jordan Peace Treaty suggests the way to move forward. Medzini and Wolf (2004, p. 203) call it "one of the most creative water treaties on record," and a former director of Jordan's Ministry of Water & Irrigation says it not only provides a sound framework for sharing the Jordan River but also Annex 2 continues to be observed even when other provisions of the Israel-Jordan Treaty are not (Haddadin 2011). However, as emphasized below, it says nothing about Palestine, and its future is in doubt; see Box 1.1.[1] Fischhendler (2008, p. 79) is likely correct when he suggests that ambiguities are deliberately introduced into the text of agreements to provide "leeway to adjust the resource allocation during a future crisis without the need to renegotiate the treaty."

Box 1.1: Annex 2 of Israel-Jordan Peace Treaty—Part 1

On 21 October 2018, King Abdullah of Jordan announced the prospective cancelation of Annex 2 of its peace treaty with Israel. That statement sent a shock wave through Israel, as this Annex, which treated all elements of a water agreement, had converted a source of conflict into a basis for cooperation. However, most analysts believe that Israel bears partial responsibility for this decision. Israel had failed to comply with a MoU signed by Israel, Jordan and the Palestinian Authority in December 2013 to implement the first phase of the Red-Dead project, nor with an agreement reached in February 2015 for building a desalination plant in Aqaba (see further in Sect. 5.3). It is not clear whether Jordan's cancellation of Annex 2 is firm or still tentative. Subsequent Israeli announcement of plans to increase significantly the amounts of water that it will pump to Jordan and Palestine (including the Gaza Strip), as well as the new drought management plan, which calls for construction of desalination plants in western Galilee and Sorek, could ameliorate the situation. Annex 2 has been too important to transboundary water management in the region for it simply to disappear.

EDITOR: The date of the Jordanian announcement has been verified through https://www.haaretz.com/israel-news/jordan-canceling-parts-of-peace-treaty-with-israel-king-abdullah-says-1.6575745

This book provides significant extension of an earlier report (Brooks and Trottier 2012) that was prepared for EcoPeace Middle East (at the time Friends of the Earth Middle East, or FoEME). Shortly thereafter an abridged version of that paper was published (Brooks et al. 2013). Both papers suggest a model for a formal water agreement between Israel and Palestine. This book brings Jordan into the model and adjusts for recent developments important for water and peace. It continues past efforts to conceive of water as a flow rather than as a stock (Trottier 2018; Trottier et al. 2019). It is also careful to distinguish between water use and water consumption, even at the conceptual level. Water use occurs when someone interacts with the flow of water for whatever purpose; the water may be changed in temperature and quality, but generally it does not disappear in that act. Water consumption occurs when water is taken out of the freshwater system. Water can be consumed in only four ways:

---

[1]Box 1.1 focuses on Annex 2 on the Israel-Jordan Peace Treaty. For a broader, Israeli perspective of the issues separating Israel and Jordan, see www.https://israelpolicyforum.org/2018/10/25/the-canary-in-the-coal-mine-for-israel-and-jordan/.

evaporation, transpiration from a plant, flow into the sea, or flow into an inaccessible aquifer. The key point is that water can be used many times before it is consumed.

Underlying this book is an assumption that, sooner rather than later, the land lying between the Jordan River on the east and the Mediterranean Sea on the west will be divided between two sovereign states: Israel and Palestine—what is generally described as the two-state solution. If the reader prefers to substitute "the future State of Palestine," or "the West Bank and Gaza Strip," it will not affect the conclusions of the book.

Apart from asserting its political position as consistent with the two-state solution, this book stands to one side of the ongoing Israeli-Palestinian conflict, which has now been ongoing for 70 years, with nearly 50 of those years including a significant Israeli occupation of Palestinian land. There is no end of publications on the conflict, but, if the reader wishes to read just a single one of modest length, we recommend Lintl (2008), which in five chapters (plus a summary) reviews Israeli positions, Palestinian positions, the role of the European Union, the role of UNWRA, and the changing nature of the Israel lobby in the United States.

1.2 Water, Geography, and Transboundary Management

The bodies of water that are essential to both Israel and Palestine are interconnected to an extent that any simple division of them into *our water* and *your water* is impossible. Some agreement for joint management of the shared water is essential, and the same is also true for water that flows along the border between Jordan on the east and Israel and Palestine on the west. Further, equity and sustainability require an approach that steps away from seeing water mainly from technical and economic perspectives— what Linton (2010) calls "modern water"—and begins to look at water from social and political perspectives. Water policy must, of course, accept hydrogeological and other physical limitations, but those limitations are insufficient to indicate how water could be used—and even less how it should be used–to satisfy changing human needs and desires. The EcoPeace Proposal responds to that shortcoming, and also recognizes the context of years of conflict and of ongoing Israeli water hegemony in the region. In addition, this proposal aims for peace defined as the absence of violence, not only the absence of war. It is therefore consistent with environmental justice, that is, equitable treatment of all peoples in the region with respect to their varied interactions with the ecology.

Finally, to avoid digressing into an area that is notably complex in both theory and practice, issues of water pricing are mentioned only in passing. Readers looking for recent analyses of water pricing in Israel may wish to review the articles by Kislev in Mcgdal et al. (2013) or Loehman and Becker (2006). For irrigation pricing, see Molle and Berkoff (2007). For much the same reason, the potential of water markets is also avoided in the text. They simply do not play much role in discussions about water management and policy–not just in Palestine, Jordan, and Israel but throughout the Middle East and North Africa—partly because private property for fresh water is almost unknown in the region. The closest approach to a market-based approach together with concern for sustainability appears in Loehman and Becker (2006).

1.3   The Global Perspective

If our work can lead to agreement on water among Israel, Jordan and Palestine, it will join a long list of other transboundary water agreements around the world. To be clear, there is no shortage of transboundary water basins in the world. McCracken and Wolf (2019) updated the Transboundary Freshwater Dispute Database (TFDD) from its 1999 Register to the present day, and they identified

310 international river basins that are shared by 150 countries and disputed areas, that cover over 47% of the Earth's land surface and include 52% of world population.

More specific to this book, yet contrary to a common impression, riparian states around the world typically prefer to cooperate over transboundary water bodies rather than fight over them (Beaumont 1994; Wolf 1998, 1999a, 2007; Gleick 2000; Kliot et al. 2001; Jägerskog 2003; Katz 2011; Wouters 2013; Abukhater 2017; Cahan 2017; Ide and Detges 2018). Even western United States, which, thanks in large part to Hollywood movies, seems to have been embroiled in conflicts over water, finds this aspect of its history now described as myth (Fleck 2016). Moreover, as Weinthal and her colleagues note (2011, 149; see also Weinthal et al. 2013), joint water resource management has "a singularly important role to play both in facilitating the rebuilding of trust following conflict and in preventing a return to conflict through creating or exacerbating existing tensions." Yoffe and Wolf (1992) reviewed a number of studies over the previous two decades that purported to find causal link between water and war, and they tried to put them into deep rest (p. 199):

> The examples most widely cited are wars between Israel and her neighbors… The only problem with these theories is a complete lack of evidence.[2]

Munther J. Haddadin (2002), who saw most of those so-called wars from a front-seat position as a Jordanian cabinet minister, corroborates Wolf fully. He is also cited in Weinthal et al. (2011, p. 149) for indicating how water agreements can support peaceful activities once conflict has ended:

> As Haddadin explains, the inclusion of the Water Annex/in the Israel-Jordan Peace Treaty/not only allowed for its successful conclusion, but also has strengthened its durability through reinforcing transparency and credibility among the parties, especially as political tensions have deteriorated in the 2000s.

Unfortunately, lack of evidence does not seem to be sufficient reason to silence the false logic of *post hoc ergo propter hoc* with regard to the wars between Israel and its Arab neighbours, including in1967.[3] The reasoning goes that, because at the

---

[2]Wolf has quoted himself with the same or similar wording in subsequent publications.

[3]Latin wording for a fallacy that means literally "after this, therefore because of this" or informally "Since event Y followed event X, event Y must have been caused by event X." (Wikipedia).

end of the war Israel was better situated with respect to water resources, therefore attainment of those resources must have been a cause of the war. It is exactly this faulty reasoning that Stephan Libiszewski (1995), who wrote one of the best but least known of the early reviews of the Arab-Israeli conflict, focused on analytically. He asked four questions: Has water been a trigger? Has water been a target? Has water been a channel? And has water been a catalyst? As will be shown in Annex B, his analysis, along with that of Medzini (1997), demonstrates clearly that water was not a cause of the 1967 war between Israel and her neighbours.

After surveying worldwide international relations from 1956 to 2006, Ide and Detges (2018) also find a link between positive water-related interactions between states in the same river basin and subsequent improvement in their diplomatic relations. Moreover, the relationship is not just an association; rather, it seems that it is success at water negotiations that leads later to diplomatic gains. However, Ide and Detges caution that the relationship is stronger for states that are not in acute conflict with each other. Based on their analysis, they consider that Israel and Jordan were not in acute conflict, but Israel and Palestine were, which helps explain why the former were able to work toward a peace treaty but the latter were not. Of course, that conclusion is simplistic as a stand-alone explanation. Israel and Jordan had no significant dispute over borders, whereas Israel and Palestine are in active dispute over Israeli settlements and outposts in the West Bank, and the power balance strongly favours Israel.

Of course, the absence of water wars does not mean that fresh water is free of conflict. Far from it! However, the conflict is mainly *intra*national rather than *inter*national. Cities that want water for domestic use can find themselves in conflict with farmers who want water for irrigation. Farmers who line their earthen irrigation canals with cement can prevent leakage but also block water seeping to the wells of a nearby village. Environmentalists who argue for leaving more water in situ to ensure continued delivery of environmental services will commonly be opposed by those who want the water for commercial use. These kinds of conflicts occur every day across the world. They need to be resolved, but the resulting issues regarding allocation of water are not essentially different from those regarding the location of new highways and other land-use choices. Key issues include comparison of benefits and costs, and impacts related to scale, to class, and to gender—and of course to protection of the environment. These are issues of environmental justice, where the victims of "slow violence" are identifiable according to categories that are far more complex than nationality alone.

In a masterful paper, Petersen-Perlman et al. (2017) summarize the global situation with respect to cooperation and conflict over transboundary watersheds. They indicate that the subject needs additional review because climate change is making precipitation less and less predictable, and that can lead to conflict in places that had not experienced it before. Much of their paper is based on Oregon State University's Transboundary Freshwater Dispute Database (TFDD), which has identified 286 surface water basins that cross international borders. (There are also more than twice as many transboundary aquifers.) Clearly, it is of paramount importance that world leaders adopt peaceful ways of jointly managing internationally shared watersheds.

Their conclusions do not differ from those of earlier writers on transboundary water as a potential site for conflict and cooperation. As the authors write (p. 106):

> We stress that building institutional capacity is the strongest method to prevent and resolve water conflicts, despite its imperfections.... Once institutional capacity is established between parties, it has been proven to be resilient over time, even as conflict was being waged over other issues.

But that does not mean that it is easy to achieve such capacity. Indeed (p. 108):

> ... transboundary water relations are more complex than individual interactions, and are often both conflictive and cooperative at the same time. Moreover, they point out that not all conflict is bad, as conflict is often the method for disputes to be addressed, and not all cooperation is good, as power imbalances are often solidified in agreements.

Finally, it is essential to deal with a subject that has frustrated water allocation decision-making for centuries before climate change was even thought of as an issue. Hellegers and Leflaive (2015) ask what makes water allocation decision-making and reform so difficult in so many places around the world. They start from some widely accepted points with which most readers of this book will agree fully (pp. 273 and 274):

> The way water is allocated between users... and within sectors... affects the overall welfare of a basin and the distribution of wealth among water users. Water allocation is especially important in regions where water is scarce and water users compete to access the water they need.

> On the basis of such case studies it becomes clear that reallocations of irrigation water are often controversial, especially when carried out by administrative decisions and without properly consulting or compensating agricultural water users.

Chapter 5 below presents abundant evidence of those sorts of problems in the West Bank, though the problem is less administrative decisions than private sector choices that constrain decision-making about water allocation to an issue of efficiency with no attention to equity. Efficiency-based water decisions are difficult enough given the number of uncertainties and intangible issues that must be considered. Allocation decisions are that much more difficult when equity among individuals and groups must also be considered, and even worse when differential power relationships are added to the mix, as they almost always are in the Jordan River basin. Put simply, what is good for one group of people may be not just bad but disastrous for another.

Hellegers and Leflaive are modest in their conclusions, but the insights are worth remembering (pp. 282–283):

> First, governments have to settle for second-best options.... The article made the case for *effective* allocation that reflects a development strategy, as a second-best option, when Pareto-optimal allocation cannot be defined or measured.... As a consequence, there is a political dimension to water allocation that needs to be factored in.... This is even more so in the context of climate change, which generates additional uncertainty in water availability and demand.

> In addition, it is not clear at which geographical scale these objectives should be made compatible. Effective allocation at the local level does not necessarily mean effective allocation at national or global levels.

This last point is particularly relevant in the West Bank, except that concern is reversed; one worries that effective allocation at the national level will come to disrupt and very possible destroy effective allocation at the local level.

1.4 Water in the Near Middle East

For most of recorded history, conflict in the Middle East has had more to do with water than with land, much less oil. Few of the main rivers in the region belong exclusively to a single state, and the need to share water has challenged each people, each civilization, each government. *Genesis*, perhaps the oldest "history" of the region, contains incident after incident when water was at issue. And more than 50 direct references to water can be found in the Koran.

Recent history is no less contentious, and has of course been particularly so in the land shared by Israelis and Palestinians. Since Great Britain's Balfour Declaration[4] and significant Jewish immigration began, a collection of studies has emerged to determine whether there was enough available water to sustain them. Not surprisingly, their conclusions tended to differ depending upon whether they were undertaken by official sources or by local Zionists.

Surprisingly, Israeli-Arab (the term is deliberately expanded from "Israel-Palestinian") conflict over water plays only a small role in the paper cited just above (Peterson-Perlman et al. 2017). Based on analysis of the TFDD, the authors wrote (p. 107):

> Using a definition of "water dispute" where water was identified as the explicit cause of military action, De Stefano et al. (2010) found 38 "acute" disputes (those involving water-related violence) between 1948 and 2008; of those, 31 were between Israel and one or more of its neighbours, with none of the violent events occurring after 1970.

To anticipate a point that will play a significant role later in this book, the absence of recent conflict over water reflects in part Israel's role as the hydro-hegemon in the region but also the wide range of cooperative activities that already exist among Israeli, Palestinian, and Jordanian water officials. There is no shared water with Egypt, and, of course, relationships with Lebanon and Syria remain strained over water as with most other issues.

---

[4]The Balfour Declaration was a public statement issued by the British government in 1917 during the First World War announcing support for the establishment of a "national home for the Jewish people" in Palestine, then an Ottoman region with a small minority Jewish population. It was composed and signed by Arthur James Balfour, a Conservative politician who served as foreign secretary from 1916 to 1919, and is probably best remembered today for the Balfour Declaration. Further detail is available at:

https://www.google.ca/search?dcr=0&source=hp&ei=46VQXdSGM8SW5gLOwqygBA&q=balfour+declaration&oq=balfo&gs_l=psy-ab.1.0.0l3j0i131j0l6.33.2197..5521...0.0..0.80.329.5......0....1..gws-wiz.TDvXeMggPio.

1.5   Organization of the Book

The second chapter of this book provides a brief discussion of water quantity and quality issues in Israel, Palestine, and the Jordanian portions of the Jordan River basin, and goes on to identify which bodies of water in that region are shared and therefore require joint management. The third chapter reviews joint water management as it emerged from the Oslo agreements, as well as selected gaps in the Oslo perspectives. The fourth chapter begins with a brief review of the deficiencies of conventional approaches to transboundary water management and goes on to describe the EcoPeace Proposal. It also presents the main challenges that have been made to the Proposal as well as responses to those challenges. The fifth chapter reviews some of the agricultural transformations in the West Bank that are linked with changing water management. Chapter Six presents four supplementary approaches to sharing transboundary water in the region: (1) the joint management of the Mountain Aquifer; (2) multinational management of the Lower Jordan River basin; (3) the proposed construction of a water conveyance system from the Gulf of Aqaba to the Dead Sea; and (4) the water-energy nexus based on renewable sources. The seventh chapter deals with the politics of moving water from something that can be discussed after a final peace agreement to something that should be discussed first and guide the remainder of the Israeli-Palestinian peace process. The chapter ends with a brief set of general conclusions.

Finally, two annexes provide base information for the arguments of this book. Annex A describes the sequence of water studies from the years of the British Mandate for Palestine, until the first few years after declaration of the State of Israel. It emphasizes their differences depending upon the perspectives of the authors and of the agencies for which they worked. Annex B picks up where Annex A leaves off and reviews analytically the role of water in the numerous conflicts between Israel and her neighbours during the first 50 years of independence. It mainly seeks to determine whether water was a cause of any of those conflicts. It also describes the role of the Water Resources Working Group of the Multilateral Track of the Middle East Process, a diplomatic effort during the 1990s that followed most of the water conflicts already discussed.

With the information base of those two annexes, this book can focus on water as an issue in Israel, Palestine, and the Jordan River basin for the most recent quarter century from about 1995 to 2020.

References

Abukhater A (2017) Water as a catalyst for peace: transboundary water management and conflict resolution. Routledge/Earthscan, London

Beaumont P (1994) The myth of water wars and the future of irrigated agriculture in the Middle East. Int J Water Resour Dev 10(1):9–21

Brooks DB, Trottier J (2012) An agreement to share water between Israelis and Palestinians: the FoEME proposal—revised version. Friends of the Earth Middle East, Amman, Bethlehem and Tel Aviv

Brooks DB, Trottier J, Doliner L (2013) changing the nature of transboundary water agreements: the israeli-palestinian case. Water Int 38(6):671–686. Available as open access at www.tandfonline.com/doi/full/10.1080/02508060.2013.810038

Cahan J (ed) (2017) Water security in the Middle East: essays in scientific and social cooperation. Anthem Press, London

De Stefano L, Edwards P, De Silva L, Wolf AT (2010) Tracking cooperation and conflict in international basins: historic and recent trends. Water Policy 12(6):871–884

Fischhendler I (2008) Ambiguities in transboundary environmental dispute resolution: the Israeli-Jordanian water agreement. J of Peace Res 43(1):79–97

Fleck J (2016) Water is for fighting over, and other myths about water in the West. Island Press, Washington, DC

Gleick PH (2000) How much water is there and whose is it? The world's stocks and flows of water and international river basins. In: Gleick PH (ed) The world's water 2000–2001: the biennial report on freshwater resources. Island Press, Washington, DC, pp 19–38

Haddadin MJ (2002) Water in the Middle East peace process. The Geographic J 168(4):324–340

Haddadin MJ (2011) Water: triggering cooperation between former enemies. Water Int 36(2):178–185

Hellegers P, Leflaive X (2015) Water allocation reform: what makes it so difficult? Water Int 40(2):273–285

Ide T, Detges A (2018) International water cooperation and environmental peace making. Glob Environ Polit 18(4):63–84

Jägerskog A (2003) Why states cooperate over shared water: the water negotiations in the Jordan River basin. Department of Water and Environmental Studies, University of Linköping, Linköping, Sweden

Katz D (2011) Hydro-political hyberbole: incentives for overemphasizing the risks of water wars. Glob Environ Polit 11(1):12–35

Kislev Y (2013) Water pricing in Israel in theory and practice. In: Megdal SB, Varady RG, Eden S (eds) Shared borders, shared waters: Israeli-Palestinian and colorado river basin water challenges. CRC Press/Balkema, Leiden, Netherlands, pp 91–104

Kliot N, Shmueli D, Shamir U (2001) Institutions for management of transboundary water resources: their nature, characteristics and shortcomings. Water Policy 3(3):229–255

Libiszewski S (1995) Water disputes in the Jordan basin region and their role in the resolution of the Arab-Israeli conflict. Center for Security Studies and Conflict Research, Swiss Federal Institute of Technology, Zurich, and Swiss Peace Foundation, Bern

Lintl P (ed) (2008) Actors in the Israeli-Palestinian conflict: interests, narratives and the reciprocal effects of the occupation. Stiftung Wissenschaft und Politik (German Institute for International and Security Affairs), SWP Research Paper 3, Berlin

Linton J (2010) What is water? The history of a modern abstraction. University of British Columbia Press, Vancouver

Loehman E, Becker N (2006) Cooperation in a hydro-geologic commons: new institutions and pricing to achieve sustainability and security. Int J Water Resour Dev 2(4):603–614

McCracken M, Wolf AT (2019) Updating the register of international river basins of the world. Int J Water Resour Dev 35(5):732–782. https://doi.org/10.1080/07900627.2019.1572497

Medzini A (1997) The river jordan: the struggle for frontiers and water: 1920–1967. PhD Dissertation, University of London, London

Medzini A, Wolf AT (2004) Towards a Middle East at peace: hidden Issues in Arab-Israeli hydropolitics. Int J Water Resour Dev 20(2):193–204

Molle F, Berkoff J (eds) (2007) Irrigation water pricing: the gap between theory and practice. CABI, Oxfordshire, UK

Petersen-Perlman JD, Veilleux JC, Wolf AT (2017) international water conflict and cooperation: challenges and opportunities. Water Int 42(2):105–120

Trottier J (2018) Harnessing the commons to govern water as a flow. Agence Français de Développement, Papiers de Recherche AFD, No. 2018–76. Paris

Trottier J, Rondier A, Perrier J (2019) Palestinians and donors playing with fire: 25 years of water projects in the West Bank. Int J Water Resour Dev. https://doi.org/10.1080/07900627.2019.1617679

Weinthal E, Troell J, Nakayama M (2011) Water and post-conflict peacebuilding: Introduction. Water Int 36(2):143–153

Weinthal E, Troell J, Nakayama M (eds) (2013) Water and post-conflict peacebuilding. Earthscan/Routledge, Milton Park, Abingdon, Oxon, UK

Wolf AT (1998) Water wars and water reality: conflict and cooperation along international waterways. Water Policy 1(2):251–265

Wolf AT (1999) 'Water wars' and water reality: conflict and cooperation along international waterways. In: Lonergan S (ed) Environmental change, adaptation, and security (NATO ASI Series Vol. 65). Kluwer Academic Press, Dordrecht, Netherlands

Wolf AT (2007) Shared waters: conflict and cooperation. Annu Rev Environ Resour 32(1):241–269

Wouters P (2013). international law—facilitatingtransboundary water cooperation. TEC Background Papers No. 17, Global Water Partnership: Stockholm

Yoffe SB, Wolf AT (1992) Water, Conflict and cooperation: geographical perspective. Camb Rev Int Aff 12(2):197–213

Chapter 2
Water Sources and Water Uses—Now and in the Future

Abstract Palestine, Jordan, and Israel are among the world's most water short regions, and, increasingly, issues of water quantity are compounded with rapidly increasing issues of water quality. Though most of the larger rivers and aquifers are shared by at least two of these nations, some are not, and this chapter presents criteria from distinguishing shared and non-shared water, as only the former are subject to international management rules. In addition, special rules are needed for the Jordan River because existing treaties neglect the position of Palestine.

The Middle East and North Africa (MENA) is the most water scarce region in the world, with 6% of the world's population yet less than 1% of the world's freshwater supply. Seventeen countries in MENA live below the water poverty line set by the United Nations (Scott 2019). Israel, Jordan, and Palestine have been living to now with water consumption that exceeds renewable availability by almost 20%. Worse yet, this overdraft figure does not take into account the large volumes of water that should be left in situ to support ecosystems, provide for fisheries, flush away wastes, etc. (Gafny et al. 2010; Katz 2011b; Safriel 2011).

Annual renewable freshwater supplies among the three countries collectively are less than 3000 million cubic meters (MCM). Distributed across a population of over 22 million, this means that the region's population has less than 150 m^3 per capita annually available for all purposes. Thus, the region as a whole and each of the countries individually must deal with chronic water scarcity.

The main purpose of this book is to envision a more efficient, a more equitable, and a more sustainable water future for all residents of Israel, Palestine, and those Jordanians living in the Jordan River basin. However, in order to conceive of the future, it is essential to understand the past.

There is no shortage of books, essays, and articles about the range of water policies that have been applied over the years by academics in Israel, Jordan, and Palestine, nor by the respective ministries of foreign affairs in each of the three states. Neither is there any shortage of books about the international aspects of water issues in the region.

With all these authors focusing on water issues in the Jordan Valley countries, this book needs not go over old ground. It is sufficient to say that early policies almost all focus on options that resulted in over-pumping and lowering of water tables, that in general there has been much more policy making than policy implementation, and that it is only in the last decade that environmental issues have been recognized as essential to all three goals of efficiency, equity and sustainability—three goals that must be balanced against one another, as well as against inefficiency, inequity, and unsustainability (Biswas and Tortajada 2005).

Two other characteristics need to be made explicit: First, in each country and at all times water policy has been an active and widely debated political issue, as it should be. Second, the tenor of academic work on water for Israelis and Palestinians began to change after about 1990. Prior to that time, most studies on water emphasized the whole region, the general scarcity of water and the difficulties of maintaining its low-income and largely agricultural populations. Such studies continue to appear, but they began to be joined in the last decade of the previous century by a surprising number of books and monographs that focused specifically on water in the Israeli-Palestinian conflict, and on the potential for cooperation over water to be linked to, and perhaps a driver for, peace in the region (Assaf et al. 1993; Kally with Fishelson 1993; Eckstein et al. 1994; Lonergan and Brooks 1994; Libiszewski 1995; Wolf 1995; Shuval 1995, 1996; Elmusa 1996)—and those are only the monographs and books; the number of articles was even greater. By 2000, it was almost the case that every issue of major international water journals had an article about water in Israel and Palestine or in the Jordan River basin.

It does remain true that the Israeli approach to mitigating the water conflict has favoured resolving water issues by technical approaches, whereas the Palestinian approach has emphasized water rights and compensation. It also remains true that the former has generally dominated formal and informal discussions (Weinthal and Marei 2002; Aggestam and Sundell-Eklund 2014). Aggestam and Sundell-Eklund also correctly emphasize the linkages between hydropolitics and peacebuilding, which is missing in many of the earlier studies. They also point out that when a "technical framing of water cooperation takes precedence, /it/, tends to ignore power asymmetry and the politics of water (p. 10). One should also incorporate the argument that, "compensation and environmental liability could be added to the water negotiations as a way to advance a water-sharing agreement" (Weinthal and Marei, p. 460).

What follows is a brief summary of water quantity, water quality, and water sharing for the peoples of Palestine, Israel, and the Jordan River basin. Figure 2.1 shows the main water bodies in Israel and Palestine, and outlines the locations of the two main ground water basins. Figure 2.2 shows the principal forms of infrastructure that move this water from source to user within Israel.

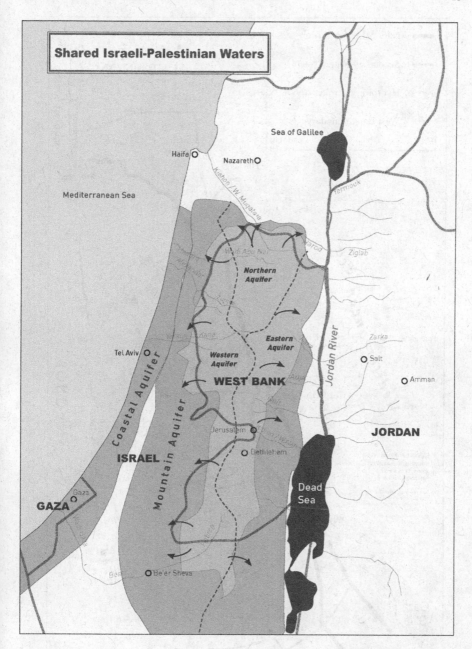

Fig. 2.1 Shared waters of Israel, Palestine, and Jordan (*Source* adapted by authors from diverse EcoPeace materials)

Fig. 2.2 Major Water Infrastructure in Israel. *Source* Feitelson and Rosenthal (2011, p. 274). *Note to editor: Permission to copy this map was received from Eran Feitelson at Hebrew University* (msfeitel@mscc.huji.ac.il)

2.1 Water Quantity: The Historic Problem

Until recently, Israel was pushing to, and often beyond, the limits of its sustainable water resources. Although desalination has provided considerable relief from concerns for drinking water, it comes with a high energy cost, as well as new environmental problems; see further in Sect. 3.4. Ironically, desalinated water is too pure for irrigation, which remains, by far, the largest use of fresh water throughout MENA. In Israel, where agriculture is a minor component of the economy, 86% of waste water is treated and re-used to account for half of Israel's irrigation (Tal 2016). (The Water Authority's goal is to reach 100% of all waste water by the year 2020.) Palestinians, too, push against and exceed the limits of sustainable water resources, particularly in the Gaza Strip (Klawitter 2007; Efron et al. 2018; World Bank 2018). However, in contrast to Israel, where agriculture is a diminishing part of the economy and where water is largely managed by a top-down modern system, agriculture remains an important part of the Palestinian economy, and water is largely managed by bottom-up community-based systems.

International statistical compilations for water are notoriously difficult because of different definitions–some countries count recycled water; others do not–and different capabilities for measurement The analysis by Reig and his colleagues (2013) at the World Resources Institute measures what proportion of total water is withdrawn every year from rivers, streams, and shallow aquifers for household, agricultural, and industrial uses. Their results indicate that Israel is even more subject to "baseline water stress" than either Jordan or Palestine.[1] However, Israel has greater financial and managerial capabilities for dealing with water stress than its neighbours.

Climate change will almost surely increase water stress throughout MENA, as indicated in numerous articles and reports, for example Sowers et al. (2011) and Kahil et al. (2015). The three countries that are the focus of this book are not likely to be exempted from the water stress that will result from climate change. As reported in a recent report (Inga 2019, p. 50):

> Over the course of the century, Jordan, Palestine, and Israel are projected to experience an average temperature rise over the Mediterranean of ~1.4 to ~4 C; a general decrease in precipitation of 25% regionally and up to 40% locally; a shift in rain seasons from winter and spring to autumn; a higher frequency and intensity of extreme weather events such as droughts, flash floods, and forest fires; as well as a growing rate of desertification.

Such effects are already evident in Israel. After five years of drought, by 2018 the Sea of Galilee (Kinneret), the largest freshwater lake in the region, had fallen to its lowest level in 17 years.[2] Even worse, its salinity had risen to its highest level in 50 years, and the Israel Water Authority had to remove 17,000 tons of salt from the lake to prevent salinity from going higher. Tal (2019) makes a persuasive case that the central driver behind the low levels in the lake in recent years involves rising

---

[1] The specific numbers indicate that 96.6% of available water is being withdrawn in Israel, 92.6% in Palestine, and 91.8% in Jordan.

[2] www.haaretz.com/israel-news/.premium-israel-s-sea-of-galilee-at-lowest-level-in-17-years-1-6632560.

temperatures and reduced rainfall rather than excessive pumping. Israel's two largest aquifers were just a few centimetres above their red lines below which infiltration from saltier water at deeper geological levels could make the water unfit for drinking.[3] Effects of increasingly extreme weather are also evident in Jordan, where in 2018 heavy rains and flash floods cost over 30 lives within two weeks.[4] Even so, Water Minister Raed Abul Saud cautioned that those rains were only 87% of the long-term annual average.[5]

Heavy rains in the winter of 2018–19 have relieved the situation for the moment, and the level of Kinneret has risen by nearly three metres, but almost everyone accepts that the future remains doubtful.[6] The last 30 years were the hottest and driest in Israeli history, and that pattern is more likely to become the norm rather than an anomaly.

About a decade ago, two scenario analyses that took climate change into account (Chenoweth 2011; Feitelson et al. 2011) found that, outside the Gaza Strip and provided that population growth is restrained, water resources in Israel and Palestine would be adequate to permit social and economic development. However, neither analysis considered that there was enough water to provide for ecosystem services. Indeed, it was not until 2004 that, in an amendment to Israel's water law, "conservation and restoration of nature" were recognized as beneficial uses of water. This change did not require that any water be allocated to ecology, but it did prevent such water from being considered as wasteful.

More recent analysis (Inga 2019) suggests that the 2011 analyses may have been overly optimistic, and that all three countries will indeed find it difficult to meet existing water commitments, much less the larger ones expected with immigration and intake of refugees in the future. The paper went on to assert that none of the three countries is building climate change sufficiently strongly into its plans for the future, nor is any building in the adaptive capacity that will promote social and political stability.

Beyond direct effects of climate change through lower and less predictable natural water supplies, secondary effects such as erosion of infrastructure and new (to the region) climate-related diseases. However, perhaps of greatest concern for the long run are effects on regional cooperation (Inga 2019, p. xx)[7]:

---

[3] www.haaretz.com/israel-news/israel-s-groundwater-levels-dropping-to-dangerous-lows-1. 6445364.

[4] https://www.bbc.com/news/world-middle-east-46161276.

[5] http://www.jordantimes.com/news/local/minister-hails-desalination-strategic-solution-water-scarcity-problem.

[6] https://www.haaretz.com/israel-news/.premium-lake-kinneret-sees-greatest-winter-water-increase-in-five-years-and-it-s-not-done-1.7084353.

[7] Here as elsewhere in this book, quotations are commonly taken from executive summaries of books or large reports because they are typically shorter and more succinct than wording in the full documents.

The bilateral agreements in place between Israel and Jordan and interim agreements between Israel and Palestine were never designed to accommodate climate-change related events such as steadily declining water availability, prolonged droughts, and other extreme weather events. The repercussions of climate change will be felt on multiple dimensions–from global outcomes to national, regional, and local reverberations. Though the writing is on the wall, especially following the Syrian civil war, the national security-related implications of climate change on the broader region's stability has not led to a significant change in policy or willingness to cooperate across borders.

Box 2.1: Excerpts from World Bank Diagnostic on Water Supply, Sanitation and Hygiene Poverty in Palestine

"In terms of poverty and water supply, Gaza and West Bank are effectively two different worlds. Gaza residents are much poorer than West Bank residents. Their access to improved water has plummeted from near total coverage two decades ago to almost zero today. More than one-third of Gaza residents are poor, as measured by the poverty headcount rate of nearly 35%. Although 95% of the population is connected to the piped network, only 1% of the population has access to improved drinking water that meets the standard of the Sustainable Development Goal indicator, according to 2016 data from the Local Government Performance Assessment. The situation has deteriorated quickly. Access 20 years ago was almost universal, and even in 2010 it was 14%. Tap water in Gaza is now undrinkable and almost everyone relies on water from small-scale desalination plants as a coping strategy.... Water provided through networked systems by formal providers is used for other domestic purposes.

"In West Bank, by contrast, poverty is much lower and most of the population uses piped water for drinking. The poverty headcount rate is much lower than in Gaza (about 16% in 2011) and access to improved drinking water is nearly universal (93% of the population). However, in West Bank, access to piped water on premises does not translate into quality services. Consider the measure of "access when needed," defined as "not available for at least one full day during the previous two weeks." By this measure, access to improved water in the West Bank drops from 93 to 80%, with the biggest decreases in the governorates of Jenin and Nablus, as well as Ramallah and Al-Bireh governorate, where service is frequently intermittent."

Source: World Bank (2018), pages xvii–xviii; references to figures and acronyms deleted.

Though the largest deficits in water supply involve agriculture, with resulting lower farm production and income, some Palestinians, particularly in the Gaza Strip, but also in some areas of the West Bank do not have access to adequate quantities or qualities of household water, commonly set as about 100 litres per person-day for drinking, cooking, and sanitation (see Box 2.1 and also Assaf 2004). In contrast, all

Israelis, with the exception of some Bedouin villages, are adequately supplied with fresh water for household use.

Finally, recent data indicate that household water consumption in Israel has been increasing—by nearly 10% in the two years ending in 2017.[8] In the past when Israel was facing stringency in water availability because of drought, households were urged to conserve water, and they did so. A particularly noteworthy achievement was made at the urban level. Israeli cities now lose less than 10% of the water that is delivered to them, a level of efficiency that Tal (2017, 134) describes as "a very low rate by international standards and an unthinkably low percentage for the Middle East."

Indeed, in the early days of desalination, some analysts argued that demand management was a cheaper alternative rather than a complement to desal (de Chatel 2007). More recently, with water supplies augmented by desalination and authorities beginning to talk about surplus, water conservation programs in Israel lagged behind those of other countries. Economist David Katz (2016) suggests a direct link between the two, which he describes as a moral hazard in Israeli water policies. After demonstrating that few consumers were even aware that water prices had decreased, which might otherwise have been an economic rationale for the increase in consumption, he writes in his conclusions:

> Water managers are always likely to seek a mix of supply and demand management strategies. Ideally, these tools would be complementary and their impact additive. This study aimed to show a potential trade-off between supply and demand-side policies, namely, the risk of moral hazard, by which consumers, knowing that additional supplies are available or are soon to be available, discount the need for conservation.

After the recent series of drought years, urban and national authorities in Israel began to regret the naïve neglect of vigorous water conservation policies for the public and for farmers. Of course, no ministry admits that its past policies were in error, but that is exactly what is implied in the following statement from the Ministry of Foreign Affairs (emphasis in last paragraph added)[9]:

> The current cumulative deficit in Israel's renewable water resources amounts to approximately 2 billion cubic meters, an amount equal to the annual consumption of the State. The deficit has also lead to the qualitative deterioration of potable aquifer water resources that have, in part, become either of brackish quality or otherwise become polluted.

> The causes of the crisis are both natural and man-made. Israel has suffered from four consecutive years of drought. The increase in demand for water for domestic uses, caused by population growth and the rising standard of living, together with the need to supply water pursuant to international undertakings have led to over-utilization of its renewable water sources.

---

[8] https://www.haaretz.com/israel-news/bad-trend-home-water-use-up-10-in-last-two-years-1. 6045731. The ensuing plan to cope with serious droughts was approved by the Israeli cabinet on 10 June 2018, but only parts of it have been released to the public.

[9] https://mfa.gov.il/MFA/IsraelExperience/AboutIsrael/Spotlight/Pages/Israel-s%20Chronic%20Water%20Problem.aspx.

The current crisis has led to the realization that a master plan for policy, institutional and operational changes is required to stabilize the situation and to improve Israel's water balance with a long-term perspective.

2.2   Water Quality: Declining Everywhere, and Rapidly

Declining water quality is a major problem throughout MENA. Most of the waste water in Israel is directed to treatment plants but less than one-third of the West Bank's population has sewage systems connected to wastewater treatment plants (WWTPs). The remainder of the population relies on septic tanks and cesspits, which are commonly poorly maintained (World Bank 2018).

Farm run-off containing fertilizers and pesticides is also common from fields across the region. The World Bank's diagnostic report is highly critical on wastewater treatment in the West Bank (2018, xx–xxi):

> In West Bank, only one-quarter of wastewater is treated, very little treated wastewater is reused, and 25 million cubic meters (MCM) of untreated sewage are discharged into the environment each year. About one-quarter of the 62 MCM of wastewater generated in West Bank is collected in sewerage networks, and two-thirds of this is treated (about 13 MCM annually). However, almost none of this treated amount is reused, due to lack of planning and to constraints on developing the necessary infrastructure to pipe the treated water to farming areas. Despite considerable investment in expanding sewerage networks, two-thirds of West Bank residents still use cesspits, which place the groundwater resource at risk of contamination.

Further, according to a report from the Arabic newspaper *Al Jazeera*,[10] in 2016 roughly 19 million cubic meters of that waste water originated from Israeli settlements built on Palestinian territory in violation of international law. Alon Cohen-Lifshitz, a researcher for the Israeli Non-Governmental Organization (NGO) BIMKOM, told *Al Jazeera* that many Israeli settlements do not have proper waste treatment facilities. To confuse the situation even more, some Israeli settlements, such as Pesagot, have their wastewater pipe directly connected to Al Bireh wastewater treatment plant. Pesagot's contribution is therefore unknown even to the Palestinian Water Authority (PWA). The waste water treated by Al Bireh's WWTP is released in the environment and mixes with untreated wastewater. Israel charges the Palestinian National Authority for all of the wastewater flowing in wadis into Israel, which in these cases includes the sewage from the settlements.

One estimate is that about 12% of settlement sewage remains untreated and travels down into streams near Palestinian communities.[11] Jerusalem does have a modern

---

[10]https://www.aljazeera.com/indepth/features/2017/09/drowning-waste-israeli-settlers-170916120027885.html.

[11]The source of the information was said to be the Knesset Research and Information Centre, but this statement could not be verified by the authors. The Centre publishes only a few of its reports, which are mainly intended for members of the Knesset.

WWTP for the western part of the city. In contrast, more than 11 MCM of waste water from the eastern portion and from Palestinian villages surrounding the city flow untreated into the West Bank. As well, the West Bank is dotted with hundreds of illegal solid waste dump sites, most of which leach into ground water. As a result, nearly all the streams that rise in the West Bank and flow into Israel are badly polluted (Alon 2002; Asaf et al. 2007; Katz and Tal 2013).

In 2017, Israeli planning officials eventually reached an agreement with the Palestinian Authority on joint efforts to improve the condition of the Kidron Valley, one of the most serious pollution hazards in the region as high amounts of raw sewage from Israeli and Palestinian communities near to Jerusalem flow through the West Bank and then into the Dead Sea.[12] Unfortunately, the project was never implemented. Two years later, the Kidron River remains probably the most badly polluted river in Israel or Palestine, with only a minor portion of the sewage undergoing treatment, and then via pipeline to irrigate date groves on Jewish settlements.

Despite the evident need to deal with existing levels of sewage, and also to permit the construction of more housing, Jerusalem's municipal water and sewage corporation seems to be engaged in a never-ending argument with the Palestinian Water Authority. It wants to pump sewage from the Kidron River to a WWTP in West Jerusalem. The PWA argues that the pipeline is just an interim solution and prefers building its own large WWTP right along the Kidron River. There is nothing new about disputes between interim/low-cost versus permanent/high-cost ways to deal with environmental problems. This one just happens to be Israeli-Palestinian and also to be tragic for many local residents.

Though generally maintaining good water quality, all three sub-basins of the Mountain Aquifer (See dashed lines in Fig. 2.1) are increasingly threatened by seepage from solid waste dumps and from sewage channels (Tagar et al. 2005; Tagar and Qumsieh 2006; World Bank 2018). Water quality in the Lower Jordan River was once good, but nearly all the good quality springs have now been diverted for local uses, and the river is seriously degraded by sewage, saline springs, and runoff from agricultural fields.

For many years, the shallow Coastal Aquifer has been polluted by runoff of agricultural chemicals, seepage from fish ponds, and seawater infiltration. About 15% of the water pumped in Israel from the Coastal Aquifer does not meet drinking water standards for chloride and nitrate concentrations. Because of a greater rate of pumping, the situation is even worse for those portions of the Coastal Aquifer that underlie the Gaza Strip. This situation was recognized nearly two decades ago (Nasser 2003; Shomar 2006), and was later confirmed by a Rand Corporation study (Efron et al. 2018) and by a joint press release from the Palestinian Central Bureau of Statistics and the Palestinian Water Authority on the occasion of World Water Day 2019;[13] see further in Sect. 3.4.

---

[12]https://www.ynetnews.com/articles/0,7340,L-4973638,00.html.

[13]http://english.wafa.ps/page.aspx?id=knUV6Va108834859056aknUV6V.

2.3 What Water Is Shared and What Is Not

Any agreement for joint management of water must be clear about exactly which bodies of water are shared. Those that are shared will be subject to any Israeli-Palestinian water agreement for joint management or Annex 2 of the Israel-Jordan Peace Treaty. Those that are not shared will be managed independently by one of the three governments.

The main water bodies west of the Jordan River are shown in Fig. 2.1. All of the dozen cross-border streams that flow to the Mediterranean are shared water, as are the three that flow to the Dead Sea or the Jordan River. One basin is doubly shared: The Hebron-Besor-Wadi Gaza basin rises in the Hebron Hills of the West Bank, then flows westward through Israel before passing into the Mediterranean in the Gaza Strip. The basin spans over 3500 km^2 and is a significant source of surface and ground water for human, agricultural, and industrial use, but it is subject to challenges ranging from pollution to excessive withdrawals (De Bruyne 2018). Rehabilitation of the northern rivers, as proposed in Israel's drought plan, should increase the volume of good-quality shared water.

The Western and Northern basins of the Mountain Aquifer are also shared water, but the Eastern Basin is deemed Palestinian, even though a few of its springs emerge in Israel. (It is useful to consider a water body as non-shared if 90% of more of its water lies within the borders of one country.) The Coastal Aquifer is not shared. According to most hydrological analyses, it is made up of a series of lenses of permeable sandstone with only limited lateral movement between them (Vengosh and Ben-Zvi 1994; see also information from the Geological Survey of Israel[14]).

In summary, roughly two-thirds of fresh water resources in Israel and Palestine can be considered as "shared water." Because the rift valley creates a barrier that blocks most east-west movement of water, only the Jordan River is shared among Israel, Jordan, and Palestine. Minor flows are mostly dealt with in Annex 2 of the Israel-Jordan Peace Treaty.

2.4 Sharing the Jordan River

Special rules are needed for sharing the Jordan River because the otherwise well designed Annex 2 of the Israel-Jordan Peace Treaty totally ignores Palestinian water rights. (Indeed, two other riparians on the Jordan River basin, Syria and Lebanon, are also ignored.) There is little interest in re-opening the Treaty, but fortunately it does "work" in a physical sense because the depth of the rift valley provides a barrier against hydrological connections between Jordan on the east and Israel and Palestine on the west.

Any thought for dividing the Jordan River's water has to begin from recognition that Jordan is one of the most water-stressed nations on earth

---

[14]http://www.gsi.gov.il/eng/?CategoryID=112&ArticleID=159.

(Scott et al. 2003; Reig et al. 2013); its allocation cannot be reduced. Therefore, the Treaty's allocation to Israel must be considered as if it had been allocated jointly to Israel and Palestine together. In the absence of any rationale to divide that portion, the EcoPeace Proposal accepts that it will be 50–50. Other divisions are of course possible without significantly changing results.

Clear borders are of course critical in dealing with transboundary water issues, but, provided the sides can negotiate with each other in good faith, it is typically not critical where those borders are located.

References

Aggestam K, Sundell-Eklund A (2014) Situating water in peacebuilding: revisiting the Middle East peace process. Water Int 39(1):10–22

Alon T (2002) Pollution in a promised land—An environmental history of Israel. University of California Press, Berkeley, California

Asaf L, Negaoker N, Tal A, Laronne J, Al Khateeb N (2007) Transboundary stream restoration in Israel and the Palestinian Authority. In: Lipchin C, Pallant E, Saranga D, Amster A (eds) Integrated water resources management and security in the Middle East. Springer, Dordrecht, Netherlands, pp 285–296

Assaf K (2004) Water as a human right: the understanding of water in Palestine. Heinrich Böll Foundation, Global Issue Papers No. 11, Berlin

Assaf K, Al Khatib N, Kally H, Shuval H (1993) A proposal for the development of a regional water master plan—prepared by a joint Israeli-Palestinian team. Israel/Palestine Center for Research and Information, Jerusalem

Biswas A, Tortajada C (2005) Appraising sustainable development: water management and environmental challenges. Oxford University Press, New Delhi

Chenoweth J (2011) Will the water resources of Israel, Palestine and Jordan remain sufficient to permit economic and social development for the foreseeable Future? Water Policy 13(3):397–410

De Bruyne C (2018) Report on the status of the Hebron-Besor-Wadi Gaza basin. EcoPeace Middle East, Amman, Bethlehem, and Tel Aviv

De Chatel F (2007) Water sheikhs and dam builders: stories and people and water in the Middle East. Transaction Publishers, New Brunswick, New Jersey, USA

Eckstein Z, Zakai D, Nachtom Y, Fishelson G (1994) The allocation of water sources between Israel, the West Bank and Gaza: An economic viewpoint. The Pinhas Sapir Center for Development, Tel Aviv University, Tel Aviv

Efron S, Fischbach JR, Blum I, Rouslan I, Karimov RI, Moore M (2018) The public health impacts of Gaza's water crisis: analysis and policy options. Rand Corporation, Santa Monica, CA

Elmusa SS (1996) Negotiating water: Israel and the Palestinians. Institute for Palestinian Studies, Beirut

Feitelson E, Rosenthal G (2011) Desalination space and power: the Ramifications of israel's changing water geography. Geoforum 43(2):272–284

Feitelson E, Tamimi AR, Bein A, Laster R, Marei A, Rosenthal G, Salhout S (2011) Defining water needs for fully exploited resources: a necessary step for Israeli-Palestinian reconciliation. Jerusalem Institute for Israel Studies, Jerusalem

Gafny S, Talozi S, Al Sheikh B, Ya'ari E (2010) Towards a living Jordan River: an environmental flows report on the rehabilitation of the Lower Jordan River. Friends of the Earth Middle East, Amman, Bethlehem, Tel Aviv: Available at: http://foeme.org/uploads/publications_publ117_1.pdf

Inga C (ed) (2019) Climate change, water security, and national security for Jordan, Palestine, and Israel. EcoPeace Middle East, Amman, Ramallah, Tel Aviv

Kahil MT, Dinar A, Albiac J (2015) Modeling water scarcity and droughts for policy: adaptation to climate change in arid and semiarid regions. J of Hydrology 522:95–109

Kally E, Fishelson G (1993) Water and peace: water resources and the Arab-Israeli peace process. Praeger. Westport, CT, USA

Katz D (2011b) Water markets and environmental flows in theory and in practice. In: Global water crisis: how can water trading be part of the solution? In: Maestu J (ed) Earthscan, London and: resources for the future. Washington, DC, pp 214–262

Katz D (2016) Undermining demand management with supply management: moral hazard in Israeli water policies. Water 8(4):159

Katz D, Tal A (2013) Rehabilitating Israel's streams and rivers. In: Becker (ed) Water policy in Israel: context, issues and options. Springer Scientific, Dordrecht, Netherlands, pp 65–82

Klawitter S (2007) Water as a human right: the understanding of water rights in Palestine. Int J Water Resour Dev 23(2):303–328

Libiszewski S (1995) Water Disputes in the Jordan basin region and their role in the resolution of the Arab-Israeli Conflict. Center for Security Studies and Conflict Research, Swiss Federal Institute of Technology, Zurich, and Swiss Peace Foundation, Bern

Lonergan SC, Brooks DB (1994) Watershed: the role of fresh water in the Israeli-Palestinian Conflict. Int Dev Res Cent, Ottawa

Nasser Y (2003) Palestinian water needs and rights in the context of past and future development. In: Daibes-Murad F (ed) Water in palestine: problems, politics, prospects. PASSIA Publications, Jerusalem, pp 85–123

Reig P, Maddocks A, Gassert F (2013) World's 36 most stressed countries. World Resources Institute, Washington, DC

Safriel UN (2011) Balancing water for people and nature. In: Garrido A, Ingram H (eds) Water for food in a changing world: Contributions from the Rosenberg international forum on water policy. Routledge, London, pp 135–170

Scott K (2019) Can the Middle East solve its water problem? Available at: https://www.cnn.com/2018/07/11/middleeast/middle-east-water/index.html

Scott CA, El-Naser H, Hagan RE, Hijazi A (2003) Facing water security in Jordan: Reuse, demand reduction, energy, and trans-boundary approaches to assure future water supplies. Water Int 28(2):209–216

Shomar B (2006) Groundwater of the Gaza Strip: Is it drinkable? Environ Geol 50(5):743–751

Shuval HI (1995) towards resolving conflicts over water between Israel and its neighbours: the Israeli-Palestinian shared use of the Mountain Aquifer as a case study. Israel Affairs 2(1):215–250

Shuval HI (1996) A water for peace plan: reaching an accommodation on the Israeli-Palestinian shared use of the Mountain Aquifer. Palest.-Isr. J 3(3–4):75–84

Sowers J, Vengosh A, Weinthal E (2011) Climate change, water resources, and the politics of adaptation in the Middle East and North Africa. Clim Change 104(3–4):599–627

Tagar Z, Qumsieh V (2006) A seeping time bomb: Pollution of the Mountain Aquifer by solid waste. Friends of the Earth Middle East, Amman, Bethlehem, Tel Aviv. Available at: http://www.globalnature.org/bausteine.net/file/showfile.aspx?downdaid=6035&domid=1011&fd=2

Tagar Z, Keinan T, Qumsieh V (2005) Sewage pollution of the Mountain Aquifer: finding solutions. Friends of the Earth Middle East, Amman, Bethlehem, Tel Aviv

Tal A (2016) Rethinking the sustainability of Israel's irrigation practices in the drylands. Water Res 90(1):387–394

Tal A (2017) The evolution of Israeli water management: the elusive search for environmental security. In: Cahan JA (ed) Water security in the Middle East: essays in scientific and social cooperation. Anthem Press, London, pp 125–144

Tal A (2019) Kinneret and climate change, letting the data tell the story. Sci Total Environ 664:1045–1051

Vengosh A, Ben-Zvi A (1994) Formation of a salt plume in the Coastal Plain Aquifer of Israel: The
    Be'er Toviyya Region. J of Hydrology 160:21–52
Weinthal E, Marei A (2002) One resource two visions: the prospects for Israeli-Palestinian water
    cooperation. Water Int 27(4):460–467
Wolf AT (1995) Hydropolitics along the Jordan River: scarce water and its impact on the Arab-Israeli
    conflict. United Nations University Press, Tokyo
World Bank Group (2018) Securing water for development in West Bank and Gaza: Sector note.
    Washington, DC
Becker (ed) Water policy in Israel: context, issues and options. Springer Scientific, Dordrecht,
    Netherlands

Chapter 3
The Existing Oslo Arrangements

Abstract Though they were supposed to be only an interim agreement with final status negotiations to be discussed within five years, and though many portions of that agreement have been ignored, the Oslo Agreements have had a lasting effect on recent and current water management in Israel and Palestine, in some cases to their benefit, in others as obstacles. After reviewing the current informal acceptance of Oslo, the chapter goes on to review two of the singular aspects of regional water management: the introduction of large-scale desalination plants in Israel; and the singularly difficult water situation in the Gaza Strip.

The Israeli-Palestinian Interim Agreement on the West Bank and the Gaza Strip, known as Oslo II, was the first to be explicit about the existence of "Palestinian water rights in the West Bank" (1995, Annex III, Appendix I), but did not define them. Oslo II did establish a framework for the management of shared water resources via Article 40. Key provisions include the establishment of a Joint Water Committee (JWC) and a Palestinian Water Authority (PWA); allocation of water between Israel and Palestine with focus on the Mountain Aquifer; and mutual obligations to treat waste water.

Though Oslo II agreements on water were favourably reviewed when they first appeared (Kliot and Shmueli 1998), over time weaknesses became apparent (Trottier 1999). Not only did it give Israel an effective veto over water management in the West Bank and Gaza Strip, but also Oslo II was intended to be an *interim* agreement governing relations between Israel and Palestine during a transitional period of not more than five years beginning in May 1994. Though the termination date is long past, both Israel and the Palestinian Authority have chosen to operate as if the water portions of the agreement were in force. However, the gaps are showing why the interim cannot continue forever.

© The Author(s), under exclusive license to Springer Nature Singapore Pte Ltd. 2020   25
D. B. Brooks et al., *Transboundary Water Issues in Israel,*
Palestine, and the Jordan River Basin, SpringerBriefs on Case Studies
of Sustainable Development, https://doi.org/10.1007/978-981-15-0252-1_3

3.1 Disputes About Failures from Continuing to Follow Article 40

Among other areas of Article 40 that indicate systemic failure are the following:

- *Do Palestinians in the West Bank have adequate water?* Some Israeli analysts claim that Israel has fulfilled its obligations under Article 40 and that the Palestinians have sufficient access to water (Gvirtzman 2012). They claim that Palestinians are under-exploiting the Eastern Basin of the Mountain Aquifer, and add that the Palestinians could increase their water supplies by treating sewage for reuse, as Israel does. Other analysts retort that Israel obstructs development of Palestinian water infrastructure (Klawitter 2007). For example, prior to the 2017 agreement on the revival of the Joint Water Committee (JWC; see below), projects for water infrastructure in the West Bank were not accepted unless Israeli settlements were connected to Palestinian wastewater treatment plants. As early as 2009, the World Bank was putting blame on both sides. It noted that little more than half of what Oslo II documents designated as "immediate needs" for the West Bank had been satisfied, and identified problems, "stemming from Israeli occupation, weakness in Palestinian planning and technical services, and lack of donor support or poorly articulated donor coordination" (2009, pp. 35, 380). The situation had only worsened when the World Bank issued its diagnostic report (2018, p. xx):

  > A combination of high non-revenue water, inadequate tariffs, and low collection rates undermine the financial viability of most service providers and impair their ability to invest in operation and maintenance. The average West Bank service provider is collecting only 76 cents on each dollar of costs. As a result, … all are dependent on public or donor finance to make capital investments or replace assets. Service providers have little room for improving services and no prospect of attracting private finance.

- *Is there over-extraction from shared water?* One of the main aims of the 1995 water agreement was to use "the water resources in a manner which will ensure sustainable use in the future, in quantity and quality" (Art. 40(3)c). Article 40 aimed to accomplish this by estimating the shared Mountain Aquifer's potential and then setting an annual baseline withdrawal rate as 483 MCM for Israel and 182 MCM for Palestine. Instead, the period since 1995 has been marked by regular Israeli over-extraction in the Western Basin of the Mountain Aquifer, in violation of Oslo II (World Bank 2009, p. 12). Back then, Israel evaded its quota by drilling into that aquifer from inside the Green Line, where the JWC has no mandate.

3.2 Bureaucratic Obstacles to Progress

Article 40 established the JWC "to deal with all water and sewage related issues in the West Bank" (Annex III, Art. 40 12). This wording excludes the Palestinians from shared management of those parts of the Mountain Aquifer that extend into

Israel (Trottier 1999; Selby 2013). It also excludes Palestinian agencies from any role with respect to the Jordan River. Though Oslo II mandates the JWC to serve as a vehicle for data sharing, fact finding, and resolution of water-related disputes, it does not indicate procedures for achieving these aims. Nor does it deal adequately with differences among Areas A, B, and C, as defined by the Accord itself (see Box 3.1).

Box 3.1: Differences among Areas A, B, and C in the Occupied West Bank

The Oslo II Accord divided the Israeli-occupied West Bank into three administrative divisions: Areas A, B and C. The distinct areas were given different statuses, according to their governance pending a final status accord: Area A is exclusively administered by the Palestinian Authority; Area B is administered by both the Palestinian Authority and Israel; and Area C, which contains the Israeli settlements, is administered by Israel. Gershon Baskin, a frequent commentator on Israeli-Palestinian affairs, described the three areas as follows:

"The divisions of the land in the West Bank were intended to have validity for an interim period of five years, initially scheduled to end in 1999… Those divisions are still valid today. Area A lands are the main Palestinian cities in the West Bank and consist of about 20% of the area (Hebron, Bethlehem, Ramallah, Nablus, Tulkarm, Jenin and Jericho). Area A is under the civil and security control of the Palestinian Authority. These are the primary urban areas of the West Bank…."

"Area B lands are the rural villages and account for about 20% of the West Bank land. Area B lands are under the civil control of the PA, but under the security control of Israel."

"Area C lands are under full Israeli civil and security control and account for about 62% of the West Bank land areas. Area C lands are the primary development areas of the West Bank for housing, economic development and… Area C is where 100% of the Israeli settlements are located and only 1%-2% of the Palestinian population. Most of the land is privately owned Palestinian land or lands designated by Israel as "state lands." Israel blocks Palestinian development in Area C and prevents Palestinian residents from building homes and infrastructure in that area.

Area C forms a contiguous territory. In contrast, under the Oslo Accords Areas A and B were subdivided into 165 separate units of land that have no territorial contiguity."

Sources: https://en.wikipedia.org/wiki/West_Bank_Areas_in_the_Oslo_II_Accord.

Baskin's comments are at https://www.jpost.com/Opinion/Encountering-Peace-The-sun-will-come-up-tomorrow-maybe-598088. Information on Israeli settlements can be found at: https://peacenow.org.il/en/settlements-watch/settlements-data/population.

To complicate the situation, the JWC must share jurisdiction over infrastructure and resource development on the West Bank with the Israeli Civil Administration (CA); see Box 3.2. According to the World Bank (2009), decision-making can take up to three years. For Palestinians, the costs in time and money of seeking CA approval for water infrastructure projects *after* they had already received approval from the JWC is highly frustrating. The inequality in power is so great that Selby (2013, p. 1) calls for "complete restructuring of Israeli-Palestinian water 'cooperation.'"A broader analysis treats the situation as a problem of hydro-hegemony and looks at the Nile River, the Jordan River, and the Tigris-Euphrates (Zeitoun and Warner 2006). Hydro-hegemony usually means that resource use is determined in ways that give the majority of the benefits to the hegemon nation, but they are not fixed and can change over time, as Zeitoun and Warner say in the abstract of their paper:

> There is evidence in each case of power asymmetries influencing an inequitable outcome – at the expense of lingering, low-intensity conflicts. It is proposed that the framework provides an analytical paradigm useful for examining the options of such powerful or hegemonized riparians and how they might move away from domination towards cooperation.

Box 3.2: Israeli Control over Water in the West Bank

Trottier (2007, p. 117) describes the process by which Israel asserted unilateral control over water in the West Bank. A few weeks after the Six Days War in August 1967, Military Order no. 92 was issued and "granted complete authority over all issues concerning water in the Occupied Territories to an Israeli officer named by the Area Commander." Those areas were not, therefore, subject to Israeli law. Subsequent military orders added specificity and, presumably, definitive authority. Notably, in December 1968 Military Order no. 291 "invalidated all prior and existing arrangements of disputes involving water."

Among other things, these military orders allowed Israel to control any new well drilling and to restrict abstraction from existing agricultural wells to the amount they had pumped in the first year of metering. However, Israel did not extend its power as far as these military orders allowed. Wherever a spring was managed by a farmer-run common property regime, Israel did not interfere with this management, apart from protecting settlers in cases when they appropriated the entire spring. Israel did not interfere either with the management of farmer-run wells that had been drilled before the occupation, apart from requiring a permit for any repair of the well. As a result, grass roots institutions that had been developed locally to manage water persisted long after the beginning of the occupation.

Yearly quotas on abstraction are often higher than the actual quantity that the well pumps for three reasons: First, growing urbanization often means that the land that had previously been irrigated from a well is now a built-up area, and household needs are far lower than irrigation needs for the same surface area. Second, many wells had been used to irrigate land that is no longer accessible to farmers since construction of the separation wall. Third, the well may have become dry or it may need repair, which requires a permit from Israeli authorities but is rarely given. Thus, the greatest limit on farmers has not been quotas, but the permit system for drilling new wells and repairing existing ones. After 1994, many unlicensed wells were drilled in the West Bank and the Gaza Strip.

For over seven years, the JWC did not officially meet as a result of the Palestinian Authority's refusal to approve Israeli water projects for settlements. This impasse left a backlog of reportedly 100 projects awaiting approval. In January 2017, Israeli and Palestinian officials announced the revival of the JWC on a somewhat different basis. Under the new arrangement, the allocation of shared natural water remains unchanged in Area C, which covers 60% of the West Bank. However, Palestinian infrastructure projects in Areas A and B, which is where most Palestinians live, can now be submitted directly to the CA without seeking initial approval from the JWC (Efron et al. 2018). However, projects that affect wells or underground water must still pass through the JWC.

The effects of the Oslo process were even more disturbing on the ground, especially for Palestinians, and are well described in Trottier's (2019) report published by PASSIA (Palestine Academic Society for the Study of International Affairs). Palestinians had always managed water communally at the local level. The Oslo Accords created the Palestinian Water Authority (PWA) and put it in charge of managing all of the water attributed to the Palestinians. In the case of the West Bank, the Oslo II Accord of 1995 allocated quantities of water to Israelis, on one hand, and Palestinians, on the other hand, from each of the three parts of the Mountain Aquifer, as distinguished by the direction of water flow (see Box 3.3). Further confusion arose because of the failure to distinguish between water use and water consumption. As indicated above, water use occurs when someone interacts with the flow of water for whatever purpose; the water may be changed in quality but by and large it does not disappear. Water consumption occurs when water is taken out of the freshwater system by evaporation, transpiration, flow into the sea, or flow into an inaccessible aquifer. The key point is that water can be used many times before it is consumed, and therefore those two terms must be carefully distinguished.

Box 3.3: Allocation of Water according to the 1995 agreement, Annex 10, Paragraph 20, Article 40 of the Protocol Concerning Civil Affairs
Source: Trottier (2019)

	Million cubic meters for Israel	Million cubic meters for PA
Western Aquifer	340	22
Eastern Aquifer	40	54 + 78 to be developed
Northeastern Aquifer	103	42

In the last few years new elements have come into play as the Palestinian Water Authority has tried to bring its urban services more closely in line with services elsewhere in the developing world. Notably, in 2009 it initiated a reform process based on the principles of integrated water resources management that "led to the enactment of a National Water and Wastewater Strategy in 2013 and a new Water Law in 2014" (Nour and Al-Saidi 2018). At least as a start, reforms seem to focus on existing urban utilities in Ramallah, Jerusalem, and Bethlehem, but, but, according to a World Bank Sector Report in 2018 (p. 8):

> The 2014 Water Law was designed to clarify accountabilities and establish autonomous utilities but implementation has been slow due to an incomplete legal structure, lack of financing, and lack of clarity of rules and responsibilities at the local level. Previous water laws and strategies have also called for the establishment of regional utilities, but there has been no progress.

3.3   No More Oslo

"No progress" is a mild way to describe not just the 2014 water law but the entire Oslo-based Israeli-Palestinian peace process. Gershon Baskin, one of the most careful observers of that process (and in some cases an actor in it), described the situation in 2019 as "No More Oslo:"[1]

> From September 1993 until September 1999, the governments of Israel and the Palestine Liberation Organization signed six agreements:
>
> - Declaration of Principles – September 1993
> - Paris Economic Protocol – April 1994
> - Agreement on the Gaza Strip and the Jericho Area – May 1994
> - Israeli-Palestinian Interim Agreement on the West Bank and the Gaza Strip – September 1995
> - Wye River Agreement – October 1998
> - Sharm el Sheikh Memorandum – September 1999

---

[1] https://www.jpost.com/Opinion/No-more-Oslo-597340.

Way before August 2019, all six of these agreements were breached in the most substantive ways – by both sides. Both Israel and the PLO in fact breached the most significant and basic obligations they took upon themselves, and did not implement what they agreed to do. The failure of the Oslo peace process is the failure of both parties.

Anticipating Baskin's almost sorrowful comment on the demise of the Oslo peace process was the report from the Israel State Comptroller, which released a report in May 2017 today on the state of transboundary water issues between Israel and the West Bank and the Gaza Strip.[2] After declaring that these issues are "the most critical and of national concern," the reports criticizes strongly the Civil Administration (local authorities in areas occupied by Israel) as well as Israeli agencies responsible for Environment, Health, and Water for failing to prevent transboundary water pollution.

3.4   The Impact of Large Volumes of Desalinated Water in Israel

Desalinated water is playing so significant a role in Israel's recent water policy that Eran Feitelson (2013) has designated it as the latest of four eras in Israeli's national water policies. (The first three were: Hydraulic Mission 1948–1964; Wise Management 1959–1990; and Reflexive Deliberation 1990–2005.) The opening of the first large-scale seawater desalination plant in Ashkelon in 2005 marked the division between the third and fourth eras.

Desalination came to prominence in Israel thanks to decreasing cost, on the one hand, and a collection of domestic and regional issues, on the other. Among the latter were increasing frequency of droughts, allowance for Jordanian and Palestinian claims on water, providing more water for environmental benefits, and declining power of the agricultural sector in Israel. Today Israel has five large desalination plants in operation, all but one of them capable of delivering more than 100 million cubic metres (MCM) of potable water per year. Together in 2018 they were able to provide 70% of Israel's household water (up from 35% just a few years ago), with some "desal" left over and mixed with treated waste water to supplement the supply of irrigation water. The goal is to be able to supply 100% of the country's drinking water and more than 40% of total needs for potable water by the year 2020.

Feitelson (2013, p. 26) describes the range of benefits from desalination:

> The advent of large-scale desalination allows for the first time since the mid-1960s to substantially augment the quantity of available freshwater. Hence, it allows for an increase in the total amount of freshwater for all sectors. Desalinated water also reduces the salinity of wastewater, thereby allowing for wider wastewater recycling (Tal 2006). Since higher wastewater treatment standards were promulgated, a wider array of crops can be irrigated by such recycled water. As wastewater is generated from the urban sector, it is not affected

---

[2]http://ecopeaceme.org/ecopeace-in-the-news/media-releases/may-16-2017-israel-state-comptroller-report-state-transboundary-water-issues-israel-palestinian-territories/. The full report is available from the website of the State of Israel Comptroller, but only in Hebrew. Informal translations into English of selected portions of the report are available from www.ecopeace.org.

by weather or climate change. Hence, the combination of desalination and higher-quality wastewater reduces the vulnerability of Israel to weather vagrancies and climate change. Yet, desalination increases energy use in the water sector, and hence its carbon footprint, and may have some detrimental effects on coastal seawater.

Lattermann and Höpner (2008) focus on the potential negative impacts on the environment of seawater desalination. They find that those impacts include discharges to the marine environment of salt concentrates and chemicals, emissions of air pollutants, and the energy demand of the various desalination processes. None of these impacts is insurmountable. Unless the plant is sited on a bay that permits recirculation of sea water, piping to the open sea will eliminate most impacts of saline concentrates, albeit at a significant cost for infrastructure. Air pollution effects are modest, especially in comparison with that generated by electrical generation for other uses, and continuing gains in the efficiency of the desal processes are reducing energy demand.

What is of concern is the absence of minerals and metals, such as magnesium, that are needed, albeit in tiny quantities, in desalinated drinking water.[3] Magnesium shortages can raise the risk of heart disease and heart attacks with studies being undertaken to examine the option to add magnesium to desalinated drinking water. It also appears that magnesium deficiency can also hurt agricultural products, and, just as with drinking water, research is underway to evaluate the potential of adding magnesium to common fertilizers.

From another perspective, Feitelson and Rosenthal (2012) describe what they call the changing water geography created by the advent of large-scale desalination. In effect, it moves the coastline inward, toward the West Bank and Jordan, which at present have no desalination facilities. As well, continued irrigation of farm land with treated waste water has its downsides. Notably, soil structure is adversely affected (Travis et al. 2010; Tal 2016). To counter this effect, Israeli farmers began to desalinate the treated waste water, which reduced the problem but also increased costs. Then in 2017, the Israel Water Authority completed a cost-benefit analysis that demonstrated that it was cheaper to desalinate treated wastewater before using it for irrigation than to use it without desalination and also use more of it to wash the salt off the land. It appears that shortly thereafter, the Authority made an official decision to desalinate treated waste water destined for irrigation.

Desalination and drip irrigation have both had a major impact on water issues in Israel, somewhat less so in Palestine and Jordan, and also elsewhere in the Middle East. However, they are not silver bullets; they are not appropriate in all cases or for all crops. See further in Box 3.4. In addition, no matter how undertaken, desalination is expensive, which explains why there is an almost perfect balance between a country's Gross Domestic Product (GDP) per capita and the proportion of its water supply that comes from non-conventional sources (Scott 2019), and why half of the world's desalination capacity is found in MENA.

---

[3]https://www.haaretz.com/israel-news/.premium-desalination-problems-begin-to-rise-to-the-surface-in-israel-1.5494726.

Box 3.4: Drip Irrigation and Desalination—Less Than Silver Bullets

As indicated at many places in the text, drip irrigation and desalination have made a major impact on thinking in the Middle East, and especially in the Jordan Valley countries. However, they cannot do everything. This box will indicate briefly one major limitation of each drip and desal.

One Major Limitation of Drip Irrigation

The earliest records we have of drip irrigation come from China and South America, but its modern history starts from the work of Simcha Blass and his son in the early 1960s who patented the first practical drip device. However, from earliest to latest times, drip systems deliver water right to the roots of plants or trees, and thus reduce waste. They cannot be used to water grains and other field crops that grow dispersed over the whole field. Other sorts of irrigation systems have to be, and have been, developed for them.

The "waste" that drip irrigation reduces was not a waste for everyone. The water that percolates through the soil via other forms of irrigation may supply the neighboring well, or may inadvertently water the neighboring field. A switch to drip irrigation means a switch in water trajectories. More water flows to the plants cultivated by the owner of the drip irrigation system and less water flows through the previous trajectories. Drip irrigation is an efficiency gain only for those who install it. It is a loss for all the users of the previous trajectories, whether human or an ecosystem.

One Major Limitation of Desalinated Water

As with drip irrigation, desalination of water has a long history, but it becomes a modern issue only with the engineering that permits construction of industrial operations that produce many thousands of cubic metres of potable water a year. The major use for desalinated water is for drinking and cooking water as well as local sanitation. It cannot be used to irrigate crops, at least not without adding some nutrients. The role of water in agriculture is to carry the nutrients to the plant or tree; it is the nutrients that promote growth. Because Israel has a current surplus of desalinated water, some of it is combined with treated waste water for use in irrigation. However, this is a short-term and rather expensive way to irrigate.

Finally, as many other writers have done, it is worth remembering Major General Abraham Tamir's cryptic comment (1988, p. 56):

> Why go to war over water? For the price of 1 week's fighting, you could build five desalination plants. No loss of life, no international pressure, and a reliable supply you don't have to defend in hostile territory.

3.5   Water for the Gaza Strip

Special attention outside the framework of an overall agreement on shared water must be devoted to water problems in the Gaza Strip. The Strip has about 2 million people living in 365 km^2, which makes it one of the most heavily populated regions in the world.

The Coastal Aquifer, which is located under the coastal plain of Israel and the Gaza Strip, is its only large source of natural water. As indicated above, most analysts accept that the Coastal Aquifer is not shared water. However, the renewable extraction potential for Gaza's portion of the Coastal Aquifer is only about 60 million cubic meters of rain water annually, and with its rapid population growth and with 86% of its water coming from municipal wells, the demand for aquifer water has surged well beyond sustainable capacity. The results of this over-pumping are highly discouraging as described by the World Bank diagnostic study (2018, p. 16):

> Sustainable renewable resources in Gaza total about 90 L per capita per day, but all of this is undrinkable and half is allocated to agriculture. Total water availability in the Gaza Strip in 2014 was 179 MCM. Of this, only one-third was sustainable groundwater yield and almost all the rest was overdraft. ... However, *all* the water is so saline that it is undrinkable. Drinking water is largely supplied by private reverse osmosis/desalination/plants, which provide more than 4 MCM to Gaza households, representing 97 percent of the water used for drinking and 67 percent of the water used for cooking...Purchases from Mekorot are also on the rise. Purchases in 2016 were about 8 MCM, compared to the 3 MCM purchased in 2014... Gaza's only internal renewable water resource, the Coastal Aquifer, which until a few years ago provided abundant fresh water to the population, has suffered extreme unregulated overuse. Effectively, groundwater abstraction is out of control. Part of the problem is that the population has responded to water scarcity with a rapid expansion of private well drilling, which PWA has been unable to regulate. In 2014, abstractions of 170 MCM were almost three times the sustainable yield.

The inevitable result of ongoing heavy withdrawals in the Gaza Strip for over a decade is steady infiltration of sea water into the Aquifer, with levels of salinity rising well beyond World Health Organisation health regulations for drinking water (Efron et al. 2018). To compound the Strip's problems, sewage collection and treatment is well below appropriate levels, and it too is seeping into what might otherwise be potable supplies.

With so many people drinking contaminated water, it is not surprising that health impacts in the Gaza Strip extend to neighbouring Israeli communities and that economic development is constrained (Weinthal et al. 2013; Shomer 2006; World Bank 2009). According to a collection of studies brought together by Israel's Institute for National Security Studies (INSS) (Kurz et al. 2018),[4] discharge of untreated sewage generated by the two million inhabitants into shallow ponds, which eventually percolates into the aquifer, has caused alarming levels of Nitrate (NO_3). Until March 2018, on most days more than 100,000 cubic meters of untreated sewage from the Gaza Strip flowed into the Mediterranean Sea not far from the Israeli town of Ashkelon.

---

[4]A summarized version of this report was published jointly by INSS and EcoPeace and is available as Bromberg et al. (2018).

Most of the problems stemmed from the inability to build new treatment facilities as a result of Israel's policy of blockade and inadequate electricity, which forced the old WWTPs to shut down. In March 2017 a new WWTP was completed by the PWA with World Bank and European support. This plant can treat all of the City of Gaza's sewage but remains vulnerable to electricity shortages (Bromberg et al. 2018).

Under the Oslo Accords, Israel committed to double the sale of water to Gaza from 5 to 10 MCM annually. However, this agreement was not fully implemented until March 2015. Because of lack of storage capacity, only 8 MCM were provided until early 2017, when the completion of some infrastructural work allowed the 10 MCM to flow into the Gaza Strip. In July 2017, Israel and the PA agreed on the sale of 10 additional MCM of water, a deal that was struck as part of a revised version of the Red Sea-Dead Sea Water Conveyance project. Once the agreement comes fully into effect, Gaza will have the option to purchase as much as 20 MCM annually from Israel.

Looking forward, it is hard to see how major change for water supply in the Gaza Strip can come without significant investment in desalination, and there have been developments toward this goal. In July 2019 WAFA (Palestine News Agency) reported that the Water Authority had completed the Gaza Strip's first sea water desalination plant project.[5] Located in Gaza City and funded by the Kuwaiti Fund through the Islamic Development Bank, the plant will have a daily capacity of 10 thousand cubic metres and, once it is in full operation, will serve upwards of 200,000 people, about 10% of the Strip's population.

In addition, in March 2018, donors pledged €456 million toward the Gaza Central Desalination Plant & Associated Works Project. The plant is expected to provide 55 MCM of water per year, with some of the needed electricity provided by solar facilities (Bromberg et al. 2018). Larger solar facilities could prosper in Gaza's sunny climate, but available land is simply not available.

For many years, untreated waste water had been purposefully infiltrated in the east of the Gaza Strip because of Israel's opposition to it being released directly into the sea. The sea currents move northwards and would have brought this pollution to Ashkelon coast where it would have harmed the operation of Israel's oldest desalination plant. As a result, untreated wastewater flowed underground from the East of the Gaza Strip towards the coast. The drinking water wells used by Gaza City, which lie in this trajectory, were going to be contaminated unless a solution was found. By 2019, the PWA had drilled the first half of an expected 28 agricultural wells to intercept the wastewater underground. The goal of this project is to pump the wastewater flowing underground and irrigate with it twice to improve its quality before it returns to the aquifer and reappears in the drinking water wells of Gaza City, a process that will prevent the health crisis that would otherwise be expected.

In addition to the project for Gaza City, some relief is also coming for residents in the northern end of the Gaza Strip. It seems that, in order to protect its own ground water, Israel decided to "once again have a pipeline connect it with Gaza—after

---

[5]http://english.wafa.ps/page.aspx?id=CpnuTsa111045781275aCpnuTs.

12 years of disengagement—and help alleviate its wastewater problem."[6] Approval has been given in Israel for construction of an underground sewer that will be 15 km long and cross the border and run through the agricultural fields to Sderot where it will connect to the sewage treatment plant that services the Sha'ar HaNegev regional council. Of course, the new pipeline will be expensive—more than US$4 million–but, as usual in other cross-border sewage projects, Israel will deduct the money from Palestinian international sales.

Also relevant is Israeli's approval of a 100 megawatt electrical line that will be funded by Qatar and that will almost double the Gaza Strip's current capacity.[7] Though intended for general supply, some of that capacity is expected to ensure that sewage and wastewater pumps are no longer shut down for lack of power.

Plans for the two infrastructure projects, plus a decision to allow an increase in the number of Palestinian day-workers who may enter Israel each day, suggests to political journalist Michael J. Koplow that, "leaving aside the short term success of these policies and questions about their durability, this change represents Israel's willingness to try something new, and… a chance to evaluate the argument that the decade-old approach of crippling blockade has been an abject failure and that some new thinking is required."[8]

The overview in the INSS review cited above cautiously summarizes the challenge facing the Gaza Strip and the ensuing dangers if collective action is not taken soon (p. 7):

> … the impact of a reconstruction project will only be visible over time. This means that awareness regarding the hardship in the Strip, formal agreement regarding the need to improve the living conditions and infrastructure there, and even the initial mobilization of resources and initiation of processes in this direction will not be enough to prevent a collapse of the Strip's social and economic infrastructure or to completely prevent the recurrence of a violent clash between Israel and Hamas, which could escalate into an all-out confrontation. Nonetheless, it is necessary to formulate the outline of a reconstruction project and to take measures, with the aim of increasing over time the chances of mitigating the risk factors originating in the Strip and establishing a period of social and security calm that presumably would enhance the motivation to invest in infrastructure in the region.

To summarize, the Gaza Strip is indeed a tough case with respect to water supply (and lots of other things too), but it is not impossible to think of the region as having adequate, good quality water sufficient for its population provided that transboundary cooperation replaces conflict, and does so for the long term.

---

[6]https://www.ynetnews.com/articles/0,7340,L-4995304,00.html.

[7]https://www.moroccoworldnews.com/2019/06/276185/qatar-makes-plans-for-gaza-power-line-further-aid-for-palestinians/.

[8]This comment appears in *Israel Policy Forum* for 18 July 2019 under the heading, "There's something happening; however, what it is ain't exactly clear.".

References

Bromberg G, Elad O, Eran O, Giordano G (2018) Gaza on the edge: the water and energy crisis in Gaza. EcoPeace Middle East, Amman, Bethlehem and Tel Aviv; and INSS, Tel Aviv

Efron S, Fischbach JR, Blum I, Rouslan I, Karimov RI, Moore M (2018) The public health impacts of Gaza's water crisis: Analysis and policy options. Rand Corporation, Santa Monica, CA

Feitelson E (2013) The four eras of Israeli water policies. In: Becker N (ed) Water policy in Israel: context, issues and options. Springer Science, Dodrecht, Netherlands, pp 15–32

Feitelson E, Rosenthal G (2012) Desalination space and power: The ramifications of Israel's changing water geography. Geoforum 43(2):272–284

Gvirtzman H (2012) The Israeli-Palestinian water conflict: an Israeli perspective. Begin-Sadat Center for Strategic Studies, Mideast Security and Policy Studies No. 94, Ramat Gan, Israel

Klawitter S (2007) Water as a human right: the understanding of water rights in Palestine. Int J Water Resour Dev 23(2):303–328

Kliot N, Shmueli D (1998) Real and ideal institutional frameworks for managing the common Arab-Israeli water resources. Water Int 23(4):216–226

Kurz A, Dekel U, Berti B (eds) (2018) The crisis of the Gaza Strip: a way out. Institute for National Security Studies, Tel Aviv, and EcoPeace Middle East, Amman, Bethlehem, Tel Aviv

Lattermann S, Höpner T (2008) Environmental impact and impact assessment of seawater desalination. Desalination 220(1–3):1–15

Nour R, Al-Saidi M (2018) Regulation reform process and perception in the Palestinian water sector. Water Int 43(7):908–925

Scott K (2019) Can the Middle East solve its water problem? Available at: https://www.cnn.com/2018/07/11/middleeast/middle-east-water/index.html

Selby J (2013) Cooperation, domination and colonisation: the Israeli-Palestinian joint water committee. Water Altern 6(1):1–24

Shomar B (2006) Groundwater of the gaza strip: is it drinkable? Environ Geol 50(5):743–751

Tal A (2006) Seeking sustainability: Israel's evolving water management strategy. Science 313(5790):1081–1084

Tal A (2016) Rethinking the sustainability of Israel's irrigation practices in the drylands. Water Res 90(1):387–394

Tamir A (1988) A soldier in search of peace: an inside look at Israel's strategy. Harper & Row, New York

Travis M, Wiel-Shafran A, Weisbrod N, Adar E, Gross A (2010) Greywater reuse for irrigation: Effect on soil properties. Sci Total Environ 408(12):2501–2508

Trottier J (1999) Hydropolitics in the West Bank and Gaza Strip. PASSIA Publications, Jerusalem

Trottier J (2007) A wall, water and power: the Israeli separation fence. Rev Int Stud 33(1):105–127

Trottier J (2019) Palestinian water management—policies and pitfalls. PASSIA Bulletin, Jerusalem

Weinthal E, Troell J, Nakayama M (eds) (2013) Water and post-conflict peacebuilding. Earthscan/Routledge, Milton Park, Abingdon, Oxon, UK

World Bank (2009) Assessment of restrictions on palestinian water sector development. Sector Note for West Bank and Gaza, Washington, DC

World Bank Group (2018) Securing water for development in West Bank and Gaza: Sector Note. Washington, DC

Zeitoun M, Warner J (2006) Hydro-hegemony—a framework for analysis of trans-boundary water conflicts. Water Policy 8(5):435–460

Chapter 4
Designing a New Water Future for Israelis and Palestinians

> *… it is safe to suggest that the number of words written about water in the Israeli–Arab context, per unit of water, is significantly higher than for any other water conflict.*
> Feitelson (2017, p. 15)

Abstract This chapter begins with what a brief description of water management and water governance in each Israel, Palestine, and Jordan, as well as a review of non-governmental and of private organizations involved with water management in the region. It then shifts attention toward a new approach for managing and governing shared water both from a conceptual perspective and then from a specific approach known as the EcoPeace Proposal, which could be applied now, in advance of a Final Status Agreement between Israel and Palestine. It is no surprise that such a proposal has come in for review and critique, the most important of which are presented and then critiqued in this chapter.

> **Box 4.1: Per Capita Water Use in Israel**
>
> …water use in Israel has increased in the past two decades. However, accounting for population growth in Israel during said years (some 2% annually, on average), water use data from the past two decades indicate a decrease in consumption per capita, though the past few years have seen an increase—from 331 cubic metres (CM) in 1993 to a low of 234 CM in 2011 followed by an increase to 257 CM per capita in 2016…. The data indicate that in 2016, per capita water use across sectors was 257 CM annually, of which consumption for domestic purposes (including public institutions) was 96 CM annually…. The period between 2008 and 2011 saw a considerable decrease in use to a low of 234 CM per capita across all sectors (a 13% drop) and 85 CM per capita in domestic use (a 17% drop). Since then, increases of 10% in overall use and 13% in domestic use were recorded to the 2016 levels.
>
> *Source*: Ido Avgor (25 Feb. 2018). *Israel Water Sector—Key Issues*. Jerusalem: The Knesset Research & Information Center, pp. 9–11.

© The Author(s), under exclusive license to Springer Nature Singapore Pte Ltd. 2020   39
D. B. Brooks et al., *Transboundary Water Issues in Israel,
Palestine, and the Jordan River Basin*, SpringerBriefs on Case Studies
of Sustainable Development, https://doi.org/10.1007/978-981-15-0252-1_4

Joint management of water is never easy, but it is particularly difficult for Israelis and Palestinians because such a high proportion of their water resources are hydro-geologically interconnected and because they have experienced so many years of conflict. They have also experienced different rates and patterns of economic development. Particularly since 1967, Israel's gross domestic product (GDP) per capita has greatly exceeded that of the West Bank and Gaza Strip. As one result, per capita household water use in Israel is now significantly higher than that in Palestine (see Box 4.1). As another, large-scale desalination has permitted Israel to reduce pumping from the aquifers (see Box 4.2). This was a major break from conditions that existed in the late 20th and early 21st centuries when steady increases in groundwater-fed irrigation in Israel (and many other places) led to an increase in pumping and a variety of generally adverse impacts on Lake Kinneret and the aquifers themselves (Feitelson and Fischhendler 2006).

Box 4.2: Sources of Fresh Water in Israel in 2017
Total production of fresh water in Israel in 2017 was 1935 billion cubic metres:
- 942 MCM were produced from wells
- 586 MCM from desalination plants
- 407 MCM from runoff.

These data mean that Israel is producing 49% of its fresh water from ground water resources and 30% from desalination plants.
Source:
http://www.water.gov.il/Hebrew/ProfessionalInfoAndData/Allocation-Consumption-and-production/20157/seker-hafaka-2017.pdf

In contrast, Palestinians remain as dependent as ever on the aquifers, mainly because they continue to depend much more on agriculture than do Israelis, both in terms of local livelihoods and as a share of GDP. Hence, the value of additional water is significantly greater to Palestinian farmers than to Israeli farmers. As farming technology improves and the economy diversifies, the total amount of water used for agriculture in Palestine can be expected to decline, at least moderately. Such a decline will certainly be regarded in official quarters as an indication of progress. Only a few MENA nations now derive even one-fifth of their GDP from agriculture; those that do are among the poorest in the region (Beaumont 2002). However, it is not clear whether "progress" so defined reflects anything beyond financial gain, nor whether that gain is distributed in a way that is socially equitable. Those issues are discussed more fully in Chap. 5 on Palestinian agriculture and the mixed effects of palm oil cultivation.

4.1 Institutions and Organizations Inside and Outside Government

The first three subsections provide brief reviews of formal and informal water management institutions in, respectively, Israel, Palestine, and Jordan. A fourth subsection identifies the more prominent non-governmental organizations (NGOs) that are active in the political dimensions of water issues. And a fifth describes private sectors activities in agriculture and agribusiness. Of course, universities and research institutes in each of the three countries are also active on water issues, but mostly from a nominally non-political research perspective. In the 1980s and 1990s, the Water Resources Working Group and the Environment Working Group, chaired respectively by the United States and Japan, both of which were parts of the "technical" multilateral track of the Middle East Peace process, were active in the region. However, in the absence of progress at the "political" bilateral track, they gradually ceased to have much influence. The activities and results of the Water Resources Working Group is reviewed in greater detail in Annex B.[1]

4.1.1 Israel

Israel has had a highly centralized, command-and-control water management system at least since 1959, when it passed a law that effectively nationalized all its water. The process that led to the situation we see today was described in detail by Feitelson (2013), but more briefly for our purposes by Feitelson and Fischhendler (2006); see also Fig. 2.2:

> After the establishment of the state it embarked on an ambitious hydraulic mission, the purpose of which was to supply water to the new settlement system built at the time to accommodate the large immigration wave and assure the state's control over its contested periphery. To this end a national water plan was prepared in 1950, on the basis of earlier plans. Its main goal was to integrate all three main sources into a single system, which will convey water from the (relatively) water abundant north to the arid south, where the greatest agricultural potential lay.

> The National Water Carrier, which connects Lake Kinneret to the northern Negev, was completed in 1964. As a result all the major aquifers were inter-connected and a single unitary system established.... From this point onward the emphasis shifted to the allocation of the existing resources. To this end a new water law was ratified in 1959.... The Water Law annulled private ownership of all water resources (both surface and underground, including also storm runoff, drainage and sewage), and placed their management with the state.... As a result of this law a uniquely centralized system was established.

Since 2007, the central agency in Israeli water policy, regulation, and management has been the Israel Water Authority. This body, which reports to the Ministry

---

[1] As will also be detailed in Annex B, David Brooks, principal author of this book, attended most of the meetings of both the Environment and the Water Resources Working Group as technical support to diplomats from Canada's Ministry of Foreign Affairs.

of Infrastructure, is also responsible for the creation, adjustment as necessary, and adoption of a Master Plan for operation of the water sector up to 2050.[2] Before 2007, water management was under the authority of the Ministry of Agriculture, with the almost inevitable result that farming received excessive amounts of irrigation water.

Perhaps the most striking part of the current Israeli water plan is that, thanks to the combination of desalination of sea water (mainly for household use) and recycling urban waste water (mainly for irrigation) extraction of water from the Sea of Galilee (Lake Kinneret) for delivery to homes, farms and businesses through the National Water Carrier has dropped by 95%—from 513 MCM (million cubic metres) in 2001–2002 to only 25 MCM in 2018–2019.[3] Moreover, plans are now underway for Mekorot (see just below) to use the same set of pipes but reverse the flow in the Carrier so that it pumps water from the desalination plants in the south of the country back to its north. At least according to the plan, it will eventually be possible to raise the level of the lake by approximately half a meter. In effect, the National Water Carrier would be using Kinneret as a reservoir to store fresh water for years when rainfall is abnormally low. Positive as this may sound, it does not come cheaply. Capital costs alone are estimated to be $278 million, and at $0.55 per cubic metre the cost of desalinated delivered to Kinneret will be several times the cost of Kinneret water delivered to the coast.

Responsible to the Israel Water Authority is Mekorot, which is a nationalized water service and delivery corporation. Mekorot supplies 80% of Israel's drinking water and 70% of its total water use. The company runs 3000 installations throughout the country for water supply, water quality, infrastructure, sewage purification, desalination, rain enhancement, etc. and operates the National Water Carrier.

Water tariffs are set by the Ministry of National Infrastructure and Ministry of Finance. They vary by use; rates for industrial and agricultural use are lower than those for household consumption and services. The bulk water tariff is the same throughout the country, regardless of the difference in supply costs.

4.1.2 Palestine

Most of the water actually used by Palestinians is still managed by local community-based or farmer-based institutions. The rest of the water is managed by municipalities. The Oslo agreements attempted to parachute Israeli-like water institutions onto the Palestinians (Trottier 2007; World Bank 2009). Few Palestinians even knew about this component of the agreements, and they continued then and still now to abide by existing grassroots institutions, which they perceive as effective and fair (Trottier 1999, 2013). As a result, the 2002 water law and 2014 water decree have never been

---

[2]Water Authority, Long-term Master Plan for the Water Sector: Part 1—Policy Paper, 4th Edition, August 2012 [Hebrew].

[3]https://www.timesofisrael.com/plan-to-pump-desalinated-water-to-sea-of-galilee-may-open-diplomatic-floodgates/.

fully implemented. At least the latter was written in Arabic; the former remained in the English in which it was composed.

The Palestinian Water Authority (PWA) was established under Presidential Decree No. 90 of 1995, which, as stated in Article 1, has independent legal personality and its own budget, along with a head appointed by the President of the Palestinian National Authority (PNA). On its website,[4] under "functions and powers of the PWA," is the following with the date 27 July 2017:

- Assume full responsibility for the management of water resources and sanitation in Palestine.
- Preparation, assuring approval and implementation of policies, strategies and water plans.
- Searching for available water sources and suggest aspects of water distribution between different sectors.
- Develop procedures and plans for the establishment and development of the company and water utilities, as well as plan and evaluate water and sanitation projects or complementary projects.
- Prepare laws, drafts and regulations concerning water and submit them for respective authorities for approval and ratification.

In effect, this broad mandate makes the PWA the Palestinian Ministry of Water Supply and Sanitation, but three other agencies share part of its responsibilities (World Bank 2018):

- The *Water Sector Regulatory Council (WSRC)* was set up under the 2014 Water Law with the mandate of sector regulator, approving tariffs, licensing and regulating service providers, and protecting consumers. However, the head of the PWA has not transferred any of these roles to the WSRC, which at present leaves it with little more than data gathering duties.
- The *Ministry of Local Government* supports, monitors, and regulates local government units, which were assigned responsibility for providing water supply and sanitation services under the 1997 Local Government Law.
- The *Ministry of Finance and Planning* allocates sector finance and manages financial flows.

Though the powers of the PWA are nominally comparable to those for the Israel Water Authority, the PWA has found it difficult to extend its authority beyond central regions of the West Bank (effectively from Ramallah to Bethlehem). Nour and Al-Saidi (2018) find that the WSRC has failed to form good working relationships with either the utilities or the PNA. In their review, they found (p. 913) that, "The overriding sense in the utilities' perception of the regulation reforms and the role of the WSRC is one of mistrust." It is equally possible that neither the utilities nor the PNA wanted any relationship with the WSRC. Clearly, the expectation that the WSRC would come into full operation by the end of 2018 was overly optimistic.

---

[4]http://www.pwa.ps/page.aspx?id=mV69sHa1941576120amV69sH.

Further, given Trottier's (1999) description of the dispersion of water supply in rural Palestine, the notion of creating rural water utilities seems even more optimistic.

Finally, as described more fully by Trottier (2019), the Oslo Accords promoted a new era where international donors took over from Israel a significant share of investment in water infrastructure in Palestine. Since 1994, in the West Bank alone, over 2000 projects concerning water have been proposed for donor funding. Almost half of them were actually funded, and 90 percent of the funded projects were implemented or ongoing in 2016. Unfortunately, underlying much of this work was the Palestinian Water Law of 2002, which had declared that water was a public good. It didn't recognize the many decentralized institutions managing springs and wells collectively as a communal property, nor did it bring any of their members into policy-making efforts within the PWA.

Palestinian farmers also have continued investing to develop new access to water in order to extend irrigation to new land, as described more fully in Sect. 5.3. Briefly, agricultural frontiers (or pioneer fronts) occur wherever agriculture is extended over previously uncultivated land, or over land that was used non-intensively. The process involves a major reconfiguration of farmers' interaction with land and water that goes beyond turning to high-value crops. Both land tenure and water tenure are commonly modified by the newly accessed supply of water.

4.1.3 Jordan

In most of Jordan, average annual precipitation is less than 200 mm. As stated by Salameh (1990, p. 69) who in the 1990s was Director of the Water Research and Study Center at the University of Jordan in Amman:

> The history of man in Jordan throughout the last three to four millennia has been determined and shaped largely by one major infrastructural element, namely water…. In Jordan the lifestyles of people, their socio-economics and their conflicts have all been determined by this basic issue.

According to Water Minister Raed Abul Saud,[5] Jordan's consumption of water for all purposes amounted to 1076 million cubic metres (MCM), including 480 for drinking water, 555 for irrigation, nearly 35 for industry, and 7 for supplying remote areas. The main water agency is the Ministry of Water and Irrigation (MOWI), which is composed of the Water Authority of Jordan and the Jordan Valley Authority (JVA), and which in 2011 prepared the National Water Management Plan (NWMP). In addition, Jordan is actively pursuing solar energy options to reduce GHG emissions and high dollar cost of fossil-fuel-generated electricity.

The MOWI has broad responsibility for socioeconomic development in the Jordanian portion of the Jordan basin, and also for water supply to urban areas. Peralta et al. (2011, p. 295) assert that, "Existing regulations and bylaws require all

---

[5]http://www.jordantimes.com/news/local/minister-hails-desalination-strategic-solution-water-scarcity-problem.

government agencies and private water developments to follow the NWMP. However, Jordanian agencies do not always do so." Further, even though MOWI has had a water demand management program since 2002, larger farmers and other agricultural interests actively oppose its policies and sometimes ignore them (Zeitoun 2009).

Nevertheless, as exemplified by the comment humorously entitled "Never let a good water crisis go to waste" (published by the Organization for Economic Cooperation and Development-OECD), Jordan's national water management officials can react swiftly and effectively when needed[6]:

> In the early 2000s, Jordan was facing a long dry spell and increasing water demand. With limited surface water supplies ....it relied primarily on rapidly depleting groundwater resources to feed its population, with agriculture as one of the key water consumers. A lack of groundwater resource protection, alongside an influx of refugees, created a water crisis that saw public demonstrations and conflicts among farmers. The government responded with a series of drastic reforms, including a Water Authority Law to stop illegal pumping of groundwater in rural areas. The law introduced jail penalties for illegal drilling, strengthened enforcement, and cut water quotas for new wells. Together, these actions resulted in a 95% reduction in illegal drilling and 30–40 million m^3 of water savings, reducing farmers' water use significantly without reducing production.

As well, Jordan always played an active role in the water-related portions of the Middle East Peace process (Haddadin 2002).

Despite all those endeavours, the water future for Jordan is far from encouraging. In what was likely the most complete review of its future, a study funded by the US Agency for International Development (USAID) (Humpel et al. 2012) nearly a decade ago looked backward and then offered a notably bleak forecast (page vii):

> Over the past 50 years, Jordan has come to depend primarily on groundwater for its municipal, industrial, and Highlands agricultural sectors. During the past 20 years, Jordan's public and private sectors have engaged in extensive well-drilling and over-pumping of groundwater that is far beyond natural recharge capacity. This over-pumping has reduced the natural base flows into the side wadis and natural springs along the rift, causing significant economic and environmental harm....
>
> Jordan is facing a future of very limited water resources ... If supply remains constant, per capita domestic consumption is projected to fall to approximately 90 m^3 per person per year by 2025, putting Jordan in the category of having an absolute water shortage that will constrain economic growth and potentially endanger public health.

As with many earlier studies this one too noted the large share of water that is taken by irrigation, the heavy subsidies required by this sector, and the small share that it contributes to the nation's GNP. Even so, at least at the start, the report cautions against quick actions to reform irrigation in Jordan (page ix):

> However, irrigation in the Jordan Valley supports a large number of jobs that would be difficult and expensive to replace, uses much of the country's reclaimed wastewater that has no other current use, is trending toward higher water use efficiency, supports export-oriented value chains, and enjoys substantial political support.

---

[6]http://www.oecd.org/agriculture/never-waste-a-good-water-crisis/
OECD (21 March 2019) by Guillaume Grure.

Those cautions are well advised, and suggest concern for more detailed reviews of which jobs might be lost and which groups of workers would be most affected. As well, there are questions as to whose irrigation water would be sacrificed in order to reduce the fiscal burden of subsidies. Further, towards the end, when the USAID report shifts from a "review of water policies" to "recommendations for strategic priorities," it offers a list that is so long and so potentially invasive as to be a near-total overhaul of the country's economic and financial management of the water sector.

4.1.4 Non-governmental Organizations Working on Water[7]

Over the years since the signing of the Oslo agreement in 1994, there have been a number of efforts to stimulate cooperation among environmentalists in Israel, Jordan and Palestine. A variety of Non-Governmental Organizations (NGOs) from the region took part in these cross-border activities during the last 25 years. However, as will be discussed at the end of this section, most of these have ceased functioning. At present, only one is still active. EcoPeace Middle East has a national director, an office, and staff in each Amman, Ramallah, and Tel Aviv. It is a strongly politically active NGO. By "politically active," we do not mean that an NGO is linked to a particular party or competes for seats in the legislature. It only means that it participates in and also generates public discussions about relevant issues.

EcoPeace's strong infrastructure has enabled it to carry out large projects connected with such environmental questions as the future of the Dead Sea and the promotion of effective management under its "Good Water Neighbours" program, which works to bring together pairs of communities located on opposite sides of a border.

Some EcoPeace efforts are deprecated by local or regional governments, which sometimes feel pushed to one side by its actions or challenged by its ability to stimulate discussion about controversial issues, such as the proposed canal between the Red Sea and the Dead Sea. On the other hand, EcoPeace has done much to introduce environmental thinking in the region, as well as to unite Israelis, Palestinians, and Jordanians behind the idea that agreements on water can serve as a key step toward peace in the Middle East.[8]

The Arava Institute for Environmental Studies serves both as a school and a research institution from its base at Kibbutz Ketura in the Negev, 65 kilometres north of Eilat. The student programs at the Institute are academically supervised by Ben Gurion University of the Negev, which is one of the fastest growing research universities in Israel. Students come from Israel, Jordan, Palestine, North America,

---

[7]We wish to thank Robin Twite of the Arava Institute at Kibbutz Ketura for assistance with the material on non-governmental organizations.

[8]See the film *Sea, Sun, and Peace?* by an American reporter named Naomi Zeveloff that came out in 2019.

and Europe, and are taught in English. Its cross-border activity operates independently and has many features of an NGO. Work has also been done by the Arava Institute to stimulate practical cooperation between Israeli and Jordanian farmers in the Arava valley, and on the provision of water treatment facilities and renewable energy to small communities in the region. Recently the Institute launched an effort to create a regional center that will deal with climate change adaptation.

An NGO which has had a long history of promoting regional cooperation is the Israel Palestine Center for Research and Information (now known as Israel-Palestine Creative Regional Initiatives). Compared with EcoPeace, which works at the implementation level, IPCRI operates at the institutional level, with participants from environmental NGOs, business leaders, and government ministries from both sides.[9] It has sponsored a number of projects and conferences on water issues.

A large number of other environmental NGOs in Israel speak out on water issues, but mainly within a broader context. The Society for the Protection of Nature in Israel, for example, emphasizes the role of water in protection of nature. The Israel Union for Environmental Defense speaks out forcefully on water when legal issues are involved. Zalul Environmental Association focuses on protecting the seas and rivers of Israel through conservation, research, awareness-raising, and education. It is active on reducing spillage and seepage of sewage and other toxic wastes into Israel's rivers.

Though there are a large number of environmental NGOs in Palestine, some of which focus on water, and a growing number of what are called civil society organizations in Jordan, most have been formed in the last few years. However, the bulk of NGOs and civil society organizations in both countries focus on social, educational, medical, and welfare issues (Karakaya 2019). At no time does Karakaya's article even refer to environmental and water NGOs in either Palestine or Jordan. This may be because in Palestine NGOs are explicitly forbidden from political activities, and in Jordan potentially forbidden though the law is often interpreted liberally (Elbayar 2005). In either case, it is hard to see environment and water as being notably more political than health and education.

Three environmental NGOs in Palestine are particularly active on water issues. The Palestinian Hydrology Group performs much like a government agency for water and sanitation issues, especially in more rural parts of the West Bank. It monitors, analyzes, and reports on the changing state of water quality, sanitation, access, and pollution. The Palestinian Agricultural Relief Committees focus on development for small-holder farms across Palestine, and, when relevant to its goals, speak out on irrigation issues. The Applied Research Institute-Jerusalem is a research establishment that focuses on natural resources, water among them, with the ultimate goal of a more sustainable Palestine.

In general, Palestinian water NGOs do not have the financial base of many individual donors that most Israeli NGOs do. As a result, they depend on international donors from other governments, which, in one way or another, circumscribes what

---

[9]http://www.civilsociety-jo.net/en/organizations/12/environmental-organizations.

they can do. NGO activities must fit donor priorities, which are not necessarily those of the communities where the NGO is working.

In conclusion, the formerly close cooperation among Israeli, Palestinian, and Jordanian environmental NGOs has declined in recent years. In part, the decline is simply a casualty of the collapse of formal peace negotiations. In part, it is also a casualty of the active movement in Jordan and in Palestine known as "anti-normalization," which discourages cooperation of any kind with Israeli organizations and individuals. Not only has this led to a decline in the willingness of a number of Arab environmental NGOs, as well as government ministries, to cooperate with their Israeli counterparts, but it also creates logistical and financial problems. Notably, scientists in the region hesitate to commit themselves to work with scientists across the border, and donors are reluctant donors to support projects that have, in their eyes, little chance of success.

Cooperation in the region over environmental problems is now very limited. Relationships over water distribution between the Governments of Israel and Jordan are maintained at a reasonable level, but other connections are not encouraged. Jordanian NGOs venturing into this area have formidable problems to overcome, both logistically and in terms of official discouragement. Relations between Israel and Palestine are even more problematic. Though their work can mitigate the adverse effects of mutual distrust and actually improve local situations, NGOs are inevitably limited in their macro impact.

4.1.5   New Private Actors on Water Issues[10]

The Oslo process was part of a broad effort at economic transformation of the region (Bouillon 2004) that assisted with the partial globalisation of the regional economy. The major consumer of water in the area, agriculture, inevitably became part of this process. Private investors think of water in terms of costs and benefits measured in monetary terms. After 1994, foreign private companies were given infrastructure contracts by donors operating in the West Bank and Gaza Strip. Mainly after 2008, they became interested in irrigated agriculture. Palestinian agribusinesses now export medicinal herbs and medjoul dates, often through contract farming. Israeli agricultural settlements in the Jordan Valley also export medjoul dates cultivated in plots located next to those of Palestinian agribusinesses. Private investors in irrigated agriculture, whether Palestinian or Israeli, all need secure access to water. At the time this manuscript is being completed, such companies are targeted by the Middle East Plan put forward by the United States; see further in Sects. 5.1 and 5.2.[11]

---

[10]This subsection focuses on the role of the private sector in agriculture and agribusiness. For a broader review of the private sector in Palestine, see Khalidi et al (2019).

[11]https://www.whitehouse.gov/wp-content/uploads/2019/06/MEP_programsandprojects.pdf.

On 25–27 June 2019, the United States gathered Israeli and Palestinian business people in Bahrain for a conference without any official Israeli or Palestinian representatives. When opening the conference, Jared Kushner, who heads the American team for a peace plan, stated, "For too long the Palestinian people have been trapped in a framework for the past. This is a framework for a brighter future. It is a vision of what is possible with peace."[12] With that start, it seems unlikely that farmer-based land and water tenure, as is common in the West Bank in 2019, was ever on the agenda for the conference. A neo-liberal outlook whereby efficiency means generating income for investors prevailed. Investors are now major actors both as companies involved in water infrastructure and in the use and, in some cases, consumption of water. This includes Palestinian agribusinesses that are currently transforming Palestinian agriculture and societal interactions with water (Trottier et al. 2019).

4.2 New Approach for Sharing Water: Conceptual Aspects

Transboundary water management is still a relatively new issue for academic analysis (Dinar and Tsur 2017), but it is an old issue for politicians. Most transboundary agreements treat water as if it were a pie to be divided among riparian states. This traditional quantitative allocation approach is currently reflected in proposals put forward by both the Israeli and the Palestinian negotiating teams (Lautze et al. 2005; Lautze and Kirshen 2009). However, though quantitative approaches to sharing water can resolve short-term issues, they have longer term defects. As indicated above, as a result of climate change, renewable water resources are likely to decrease in the Middle East, with particularly severe effects on agriculture (Freimuth et al. 2007; FAO 2008; Sowers et al. 2011; Inga 2019). Quantitative allocations that are possible today may be impossible in a few years simply because of climate change. Further, demographic change and economic development affect demands for water in ways that are at least partially unforeseeable.

Quantitative division of available water also promotes securitization. That is, water is portrayed as such an essential component of national security that it leaves the realm of what is negotiable and open to compromise (Trottier 1999, 2008; Zeitoun 2007; Feitelson et al. 2012). To compound the problem, Fischhendler (2015) emphasizes that, on the one hand, transboundary water resources are notably susceptible to claims of securitization, but, on the other, the term is often left undefined and absent of specific drivers. Given this start, it is not surprising that he also charges that de-securitization of water sources is assumed to promote economic growth without any verification.

Fischhendler is no doubt correct in his depiction of de-securitization, but he may not be taking into account a broader concept of securitization. Alatout (2007, 2008) shows how the concept of securitizing water played a crucial role initially in Zionist

---

[12]https://www.haaretz.com/middle-east-news/.premium-in-bahrain-air-of-israeli-arab-normalization-and-a-message-to-iran-1.7410754.

proposals during the British Mandate (see Annex A) and then in Israeli water policy from the 1950s when Israel's first water laws were passed. Among other things, the 2008 article (p. 4) notes that "the struggle over the notions of water abundance and scarcity was an essential part of working through the political conflicts over the meaning of Jewish subjectivity, the boundaries of the state, and its right to intervene in civil society."

Starting from explicit indications of differences as to whether water in the future state of Israel is scarce or abundant, Alatout (2007, p. 201) dissects the debate and converts its conclusions from hydrological to political and even racial:

> The implications for the debate about water scarcity are great. From this perspective, what we take as a scientific "fact" cannot be seen as an unfiltered representation of nature. Fact can be, and often is, as much a statement about politics as it is about nature.... I argue that the shift from a notion of abundance to one of scarcity in the Israeli context was a sign of another shift that was being promoted by water research and practice: a shift in the conception of Jewish identity in historic Israel-Palestine from the Jew as settler to the Jew as a citizen of the modern state.... Serving as a background to all these shifts–indeed, in some ways constituting their basic reference–was the negation of the Palestinians as a national community with political economic, social and cultural rights to Palestine's water resources.

Debate about the early nature of water policy in Israel to one side, it remains true that, though water is of course a component of one nation's security, it is equally a component of the security of neighbouring nations. Cooperation and joint management over water must become priorities on *each* state's national security agenda. Indeed, recent events in the region make it clear that national security concerns are interlocking, and therefore your security concerns are also your neighbour's security concerns.

Fortunately, even if the debate about water in the region is commonly discussed within a security framework, at the political level both sides have worked hard to keep water out of ongoing conflicts. Coskum (2009, p. 111) suggests that, though the availability of institutions to deal with water-sharing between Israelis and Palestinians has yielded a conflict *management* mechanism and encouraged water experts on both sides to continue to work together, it has at the same time "hindered the development of conflict *resolution* efforts." He also suggests (*ibid.*):

> In the absence of governmental level support, Israeli and Palestinian environmental NGOs and water experts have developed systems and infrastructure to address the water-related issues that negatively affect the quality of Israeli and Palestinian livelihoods.... /However/In spite of the effectiveness of the NGOs in addressing localised, relatively small-scale problems arising from the mismanagement of water resources, they are still far from addressing the macro-level structural issues in Israeli-Palestinian water management....

In their 2012 study, Feitelson, Tamimi and Rosenthal studied the "the potential interactions between climate change and conflict in the Israeli-Palestinian case" (p. 435). Though the scenarios they built may now be dated, their conclusions are still sound. Though water is the main issue that may be affected and /though/it does have transboundary implications:

> it is unlikely that climate change will directly affect the conflict. However, framing water as a security issue, along with the potential for furthering such securitization with reference to

climate change, may adversely affect the readiness of the parties to take adaptive measures and lead them to rigidify their negotiating positions.

From a conceptual perspective, sustainable development has been adopted as a goal by almost everyone dealing with water in the Middle East and North Africa, which is one indication that it is more a slogan than a measurable goal (Biswas and Tortajada 2005). Though some analysts think of sustainable development as a flawed concept, others try to rescue it by suggesting that it reflects a broader concept of ecological economics that is critical of the domination of neoclassical economics and that emphasizes the need for pluralism. For example, Söderbaum (2005, p. 88) argues that sustainable development is:

- Understood and measured in multi-dimensional terms incorporating social and cultural elements as well as physical, ecological, and financial ones
- Built on ethical principles that pay attention to future generations and nations other than one's own
- Built on a precautionary principle to avoid irreversible damage to people and ecosystems
- Built on normal ideas about democracy, such as participation and open access to information.

A few years later, Söderbaum (2009, p. 432) applies those concepts specifically to water management, and argues more broadly that, "Experiences from humanities and social sciences are relevant for water policy and management."

Those criteria do not make sustainable development any easier to deal with analytically, but they bring out aspects that demonstrate why "modern water," to adopt Linton's term (2010), may be economically efficient but not sustainable.

Somewhat the same approach was put forward by Kay and Mitchell (2000) when they argued for thinking of sustainability as a process rather than as a measurable goal. They then used their approach to explore the full range of opportunities for policymaking and capacity building in Israel's 1988 masterplan for water management. Among other things, they re-emphasized that, given the hydro-geology of the region, simplistic quantitative divisions of water resources into our water and their water will never be adequate for long. Further, the impossibility of trying to write general criteria for sustainable development of water resources does not preclude the ability to prepare criteria or scenarios for specific water bodies or water courses (for example, Haasnoot et al. 2012; Ge et al. 2018).

Staddon and Scott (2018) reach a similar conclusion in their editorial introduction to a special issue of *Water International* entitled "The Global Water Security Challenge (p. 1022):

Water security, set as a target (a condition to be achieved) and a process (a capacity to be developed), serves multiple institutional interests. Yet, the conceptual appeal of water security must be tempered by limits to its application in operational terms. What does it mean to be water secure? Having achieved a state or condition of water security (if this is indeed definable), what constitutes a loss or reduction of water security? …. Reflecting on these and related *what-comes-next?* questions, the articles in this issue underscore our conviction that water security will continue to evolve, taking on new dimensions of social and environmental justice, access and outcomes as well as offering a forward-looking perspective on the central role of water in broader environmental governance.

Finally, joint management of water shared by Israelis and Palestinians must accept the sharp seasonal and spatial variations in rainfall that are typical of semi-arid regions. However, what really bedevils water planning and management in semi-arid regions is year-to-year variation. Jordan, Israel, and Palestine are all subject to frequent droughts, periodic "good" years of above-average rainfall, and occasional intense storms and flooding. Sound planning of water management must therefore focus on extremes and risk minimization, not on averages and maximum utilization.

4.3   New Approach for Sharing Water: The EcoPeace Proposal

The essence of the EcoPeace Proposal is to recognise water as a flow and then to use continuous monitoring and ongoing mediation as the main management tools to achieve equity, efficiency and sustainability. These tools provide the basis for decisions to adjust withdrawals from each well or reservoir, or to modify use of water from a spring. They also encourage interaction between state and non-state actors. Ongoing mediation means that rulings or regulations can be appealed by any actor involved, whether scientist, officer of a non-governmental organization, or member of an agency that manages water. Social and economic developments over time can be accommodated and integrated within geologic, hydraulic, and engineering constraints.

As mutually interdependent riparian states, Israel, Jordan and Palestine must have the right to access and use water from shared supplies. They must also accept the parallel responsibility to maintain the quality and quantity of flow in all shared natural water sources, within the limits set (and sometimes changed) by natural conditions. Equality in rights and responsibilities does not mean that each party can expect to receive an equal volume of water. It does mean that each party will have equal standing within each of the organizations for joint management of shared water bodies. Water issues between Israel and Jordan are at present managed by Annex 2 of the Israel-Jordan Peace Treaty (see Box 4.3), but, as mentioned above, Annex 2 says not a word about Palestine.

Fortunately, the Palestinian National Authority was equally aware of the potential for using water as a route toward peace-making. According to Coskum (2009, p. 103), "Karen Assaf from the PNA Ministry of Planning underlined the importance of overcoming the lack of trust between the two sides when dealing with water-related problems by saying that 'there is a problem of conflicting entities and the attitude over the years that either we use it (water) or lose it. In essence, as Palestinians and Israelis, we have to get over this lack of trust and begin to coordinate and work positively.'"

Several earlier proposals recognized that the "most critical step towards conflict resolution is separating the concepts of territorial sovereignty from water security" (Medzini and Wolf (2004, p. 193), as well as encouraging greater transparency in water data across boundaries" (*Ibid.*). Loehman and Becker (2006) propose a regional utility and a joint commission, with the use of pricing as a tool determined by supply and demand once sustainability limits are established for all sources. However, their proposal does not take account of the dramatic differences between Israeli and Palestinian water management practices. Useful comparisons between Israeli-Jordanian-Palestinian water issues in the Jordan River basin and those in the Colorado River basin in western United States appear in a book edited by Megdal et al. (2013).

Box 4.3: Annex 2 of Israel-Jordan Peace Treaty—Part 2

The Israeli-Jordanian Peace Treaty has a number of quite notable provisions, as indicated by this quotation from Lonergan and Brooks (1994, 271):

> Article 6 of the Treaty is entitled simply 'Water.' It is devoted 'to delivering a comprehensive and lasting settlement of all the water problems between the Parties.' As such, it constitutes the first such agreement between Israel and any of its neighbours.

And by these examples cited by Kliot and Shmueli (1998, p. 218) with reference to Annex 2 of the Treaty:

> There is great emphasis on mutualism in the Treaty's provisions, namely that both Israel and Jordan need not make concessions to the other party without receiving something in exchange. For instance, according to the Agreement, Israel has to provide Jordan with water in the north .... and will receive in return groundwater in the south. Also, there is a clear-cut attempt to preserve the current patterns of use of both parties. However, because Jordan cannot store and deliver winter flows, Israel will store the much-needed water during the winter season and deliver it to Jordan in the summer.

What we believe to be the first full proposal for as general water agreement originates from Assaf and her colleagues at IPCRI (Israel-Palestine Center for Research and Information) (1993). Shortly thereafter are a series of studies undertaken to propose a joint management structure for the Mountain Aquifer (Feitelson and Haddad 1998a, b, 2000), which itself has a conceptual antecedent in a 1995 proposal by the late Hillel Shuval (1995). This work is described more fully in Sect. 6.1.

Kliot and Shmueli (1998) took a different approach. First they reviewed two documents:

- Annex Two of the Israel-Jordan Peace Treaty signed in 1994
- Agreement on Water and Sewage that is part of the Palestinian-Israeli Interim Peace Agreement signed in 1995.

Then, under the auspices of, and with funding from, the Institute for Water Resources at the Technion in Haifa, they presented "23 experts in the area of water resources management and law" with the objective of comparing the "institutional frameworks" in those two key documents with an "ideal" form of institutional framework (p. 218). Kliot and Shmueli wrote in the abstract of their paper (p. 216), "The main finding is that in most respects, expert findings did not differ significantly from the Israeli-Jordanian Peace Treaty and the Israeli-Palestinian Interim Agreement." Even more remarkably, Kliot and Shmueli found broad agreement on "issues and structures" (p. 223), including:

- Acknowledgement of a legal right of the other party to a share of common water resources
- Joint management of common water resources
- Recognition that both quantity and quality of water are extremely important
- Adherence to legal principles for management of international rivers.

Each of those principles, with the partial exception of the last one, has been central to the design of the EcoPeace Proposal from its earliest published form in 2010.

Not long after the Kliot-Shmeli paper, Feitelson and Fischhendler (2006) provided a detailed discussion of appropriate management of aquifers to ensure sustainability of both quantitative and quality characteristics. Though not initially described with reference to the Middle East, conditions in Israel were their primary case study. First, they established that the growing use of aquifers was increasing costs for water users, mainly farmers, and also imposing externalities (costs imposed on other actors than those benefiting from the pumped water), such as increased energy needed to extract water from great depths, salinization of the aquifers, and in some cases dessication. These problems call for more integrated joint management, as described in their essay:

> In many settings groundwater use has ramifications for surface water, and vice versa. These ramifications can be a result of the physical (hydraulic) connections between surface water and groundwater, or the pecuniary connections between them. That is, in many cases surface and groundwater are seen as alternative sources, whereby the increased use of one may reduce demand for the second. Hence, it has been increasingly recognized that groundwater should not be managed separately from surface water.

Feitelson and Fischhendler are cautious in their conclusions, and refrain from specific recommendations. They do recognize the need for some centralized allocation powers, and for those powers to be located elsewhere than in the Ministry of Agriculture. (As was described above in Sect. 4.1.1, this recommendation came to be accepted a year later in 2007 when the Israel Water Authority was created and commenced reporting to the Ministry of Infrastructure.) Some of their general observations are worth quoting:

> The Israeli case shows that even when a water manager is provided with exceptional power and authority – full control over all water resources and highly sophisticated management tools – he will inevitably be constrained by other governing bodies and in some cases by international treaties. Hence the decisions regarding the abstractions from an aquifer, and the use of land above it (and therefore the protection of its recharge areas) will be affected by a host of socio-political considerations, by multiple agents…

> Arguably, the concentration of power in a single institution, allows strong interest groups to co-opt this institution to their benefit, which has happened in Israeli during the term of several

Water Commissioners who came from the agricultural sector…. Rather, we propose that a
checks and balance system should be considered, where the interests of different groups and
different locales will be played out.

It may be that those of us who were the lead analysts involved with design of the
EcoPeace Proposal (*cf.* Brooks and Trottier 2012) flatter ourselves, but we believe
that it provides many of the checks and balances that Feitelson and Fischhendler
believe are necessary for sound management of ground water and surface water. See
just below in Sect. 4.4.

Finally, but by no means least in importance, to stay within sustainable limits of
their surface and ground water resources, the main focus of water management for
Israelis, Jordanians, and Palestinians must shift from supply management to demand
management (Brooks 2006). Only two of the early studies on the Israeli-Palestinian
conflict even mention demand management (Lonergan and Brooks 1994; Libiszewski
1995), and not many in the first decade of the 21st century either; Tal (2006) is a
notable exception.

This gap in thinking about the demand side of water management should not be
surprising. Demand management is still a rare concept throughout the Middle East
and North Africa (MENA) (Brooks et al. 2007), and only recently has it come to
the fore in Palestine, Jordan, and Israel. Water managers in the future must spend at
least as much effort finding ways to reduce the demand for water as they now spend
finding new sources of supply. A study by Rosenthal and Katz (2010) found many
water conservation measures were available at less than the cost of new desalination
capacity for equal volumes of water. As shown by studies in other parts of the world,
the potential for cost-effective water conservation measures is large, even if changes
in crops grown and products manufactured are ignored (see, as one example among
many, Vickers 2001). And, of course, the potential is that much larger if subsidies
available mainly for agricultural and industrial uses of water in Israel are reduced
enough to permit pricing to influence farming and manufacturing decisions.

4.4 New Joint Water Management Organizations for Israel and Palestine

Figure 4.1 shows key elements of the organizational structure for implementation of
the EcoPeace Proposal.

Two senior bodies guide decision-making in the EcoPeace Proposal: a Bilateral Water Commission (BWC) and a Water Mediation Board (WMB). The BWC
replaces today's Joint Water Committee (JWC) and eliminates the need for any further approvals by the Civil Administration. It will have responsibility for all shared
water (not just Palestinian water, as with today's JWC). The BWC makes key decisions on rates of extraction and delivery of water and on the removal and treatment
of waste water. Its decisions are based on advice from a subsidiary body, the Office

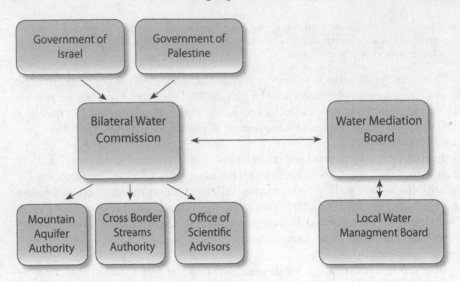

Fig. 4.1 Organigram of the Joint Management bodies in the EcoPeace proposal. *Source* Brooks and Trottier (2012)

of Scientific Advisors, made up of staff appointed or seconded by the two governments. Should the BWC find itself unable to accept a recommendation of the Scientific Advisors, or should any group or community wish to oppose a decision, the WMB can take action. The WMB, which is independent of political oversight, has a wide range of tools for resolving issues, which range from scientific investigations to public forums. The WMB also receives advice from another subsidiary body, the Local Water Management Board, which represents local bodies in reviews before the WMB.

The back-and-forth process continues until the BWC receives a recommendation from the WMB that it can accept. With this combination of forces, the EcoPeace Proposal should be flexible enough to deal with droughts that Medzini and Wolf (2004) indicate have threatened earlier water agreements. Elaboration of the manner by which the BWC and WMB will operate to promote integration of science and democracy appears elsewhere (Trottier and Brooks 2013).

4.5 Challenges and Responses

To no-one's surprise, the EcoPeace Proposal has received numerous criticisms about this or that provision. This section does not deal with specific criticism but focuses on challenges of the proposal as a whole.

The most common challenge to the EcoPeace Proposal is that it is not fully formed. Though no one can deny that further work is needed to convert the concepts and the organizations outlined above into real processes and real agencies, most of them have been tried elsewhere. Only the WMB introduces more than a modest amount of innovation. From a practical perspective, this challenge can be partly answered by suggesting a staged approach to implementation of the EcoPeace Proposal. For example, implementation might begin with a Mountain Aquifer Authority, as proposed by Feitelson and Haddad (1998a, b, 2000; see further in 6.1). Al-Sa'ed (2010) argues for a similar step-by-step proposal for wastewater treatment.

Another challenge claims that the Proposal requires reductions in Israeli withdrawals of water that would be "quite unprecedented" (Lautze and Kirshen 2009, p. 201). It is true that nations have seldom voluntarily reduced their withdrawals after establishing "prior use" of that water. However, much of the Israeli increase in water use since 1967 comes from occupied Palestinian land. This water cannot be considered as "prior use" in the usual sense. Further, the availability of large volumes of "desal" allows the Israelis to give up previously used water voluntarily. As indicated above, Israel has been able to reduce its withdrawal from the aquifers significantly because a high and growing percentage of its household water now comes from desalination. More generally, if the Israelis do have to give up sizeable quantities of water, they can expect in return to have a better quality of water flowing back to them than was formerly the case.

The burden of any cutbacks in Israeli water use would almost surely fall mainly on the agricultural sector (Lithwick 2000; Jägerskog 2003; Lautze and Kirshen 2009), which, as noted above, is a diminishing part of the Israeli economy. Israel is a sophisticated society that can accommodate the ensuing trade-offs and, if necessary, provide temporary subsidies to adversely affected farmers. Indeed, this is exactly what happened to Israeli farmers who had been living in the Gaza Strip and had to move back to Israel when Israel returned the Strip to Palestinian control in 2005.

Further, prior use as an absolute criterion for rights to water is under question, even in western Canada and the USA where it goes by the acronym of FITFIR (First In Time/First In Right). As stated by Brandes and Curran (2016, p. 45), alternatives to prior allocation provide, "the focus to better understand Canadian western water law and to identify characteristics of an emerging regime based on partnership and with an explicit emphasis on protecting water for nature." Fleck (2016) uses the theory of common pool resources to reach similar conclusions for the American Southwest. Even back in 1998, only four of the 23 experts who responded to the Kliot and Shmueli survey suggested that prior use should be a criterion for allocation of Israeli-Palestinian or Israeli-Jordanian water.

The late Shuval (2011) argued that the EcoPeace Proposal takes reallocation of the shared resources out of the hands of the two national governments and deprives both Israel and Palestine of elements of sovereignty. Although Professor Shuval's argument may be literally true, it implies that Israel and Palestine can each "eat its cake and have it too." States cannot have full sovereignty over a resource that is fluid and that moves from one state to another, from underground to surface, and from atmosphere to surface, and back again. Allocation of water is a sensitive political and legal issue, but the very concept of joint management becomes meaningless if it starts from a premise that all existing laws must remain in place.

Of course, even if a peace agreement between Israel and Palestine is reached, water will continue to be seen as a national security issue by both sides. Deadlock could then arise on the BWC with all decisions pushed over to the WMB, which is supposed to be an instrument for mediation, not arbitration. To alleviate this challenge, initial decisions should not be made at the state-to-state level, but at lower levels where the issues that typically arise (such as priority to household water) are less polemical for both Palestinians and Israelis. If necessary, the WMB will be able to call upon scientific expertise to dispel the prevalent idea that water problems are caused by "the other side."

Few of the critics of the EcoPeace Proposal saw that its greatest threat might come from the very Israeli or Palestinian officials to whom it reports. As with any agreement, either side has the power to appoint officials who will act in ways that make joint bodies dysfunctional. Some might simply not want any ongoing cooperation, or, perhaps more likely, are fearful of what might emerge from negotiations. Subramanian et al. (2012) state that barriers or, as they prefer, "risks" to reaching agreement on shared water most often stem from the perception that cooperation may expose the country to harm, jeopardize something of value to the country, or threaten the political future of individual policymakers. In the case of Israelis and Palestinian water officials, the most important barriers seem likely to fall into two categories:

- Sovereignty and autonomy—the Palestinian perception that they have less negotiating capacity and less information than the Israelis; and the Israeli perception that cooperation would intrude into its authority to make independent decisions about national development.
- Accountability and voice—the Israeli perception that the Palestinians or the regional institution may not deliver the anticipated benefits; and the Palestinian perception that their interests would not be adequately considered by the Israelis or the regional institution.

Those barriers are real, and perceptions can of course turn possibilities into probabilities. All one can say is that perceptions need not dominate the results, and that the more cooperation one experiences with people on the other side of the table, the less likely are they to do so.

In summary, there is no existing model of a transboundary water agreement that is quite like the EcoPeace Proposal for Israel and Palestine. One institution that comes close is the International Joint Commission (IJC) that has considerable power to influence management of the lakes and rivers that form much of the long border between the United States and Canada. The IJC is based on the Boundary Waters Treaty (BWT), which is now more than a century old. As Grover and Krantzberg write (2015, p. 183):

> Although the BWT and its related institutions have faced their own challenges, they have survived for over a century alongside the emergence of new treaties, agreements and institutions. In a way, the governance system has evolved over a period of time in response to the ecological, social and economic stresses.

Without arguing for any geographical analogy between transboundary waters in the United States and Canada, and those in Israel, Palestine, and Jordan, there is a modest institutional analogy between the IJC and the BWC. Each can learn from the other.

References

Alatout ST (2007) From water abundance to water scarcity (1936–1959): A "fluid" history of Jewish subjectivity in historic Palestine and Israel. In: Sandy Sufian S, LeVine M (eds) Reapproaching borders: new perspectives on the study of Israel-Palestine. Bowman and Littlefield, Lanham, MD, USA, pp 199–219

Alatout ST (2008) Locating the fragments of the state and their limits: water policymaking In Israel during the 1950s. Israel Policy Forum 25(1):40–65

Assaf K, Al Khatib N, Kally H, Shuval H (1993) A proposal for the development of a regional water master plan—prepared by a joint Israeli-Palestinian Team. Israel/Palestine Center for Research and Information, Jerusalem

Beaumont P (2002) Water policies for the Middle East in the 21st Century: the new economic realities. Int J Water Resour Dev 18(2):315–334

Biswas A, Tortajada C (2005) Appraising sustainable development: water management and environmental challenges. Oxford University Press, New Delhi

Bouillon ME (2004) The peace business: money and power in the Palestine-Israel conflict. I. B. Tauris, London

Brandes OM, Curran D (2016) Changing currents: a case study in the evolution of water law in western Canada. In: Renzetti S, Dupont DP (eds) Water policy and governance in Canada. Springer, New York, pp 45–67

Brooks DB (2006) An operational definition of water demand management. Int J Water Resour Dev 22(4):521–528

Brooks DB, Trottier J (2012) An agreement to share water between Israelis and Palestinians: the FoEME Proposal—revised version. Friends of the Earth Middle East, Amman, Bethlehem and Tel Aviv

Brooks DB, Thompson L, El Fattal L (2007) Water demand management in the Middle East and North Africa: observations from the IDRC forums and lessons for the future. Water Int 32(2):193–204

Coskum BB (2009) Cooperation over water resources as a tool for desecuritisation: The Israeli-Palestinian environmental NGOs as desecuritising actors. Eur J Econ Polit Stud 2(2):97–115

Dinar A, Tsur Y (eds) (2017) Management of transboundary water resources under scarcity. World Scientific, Singapore

Elbayar K (2005) Middle East NGO laws in selected Arab states. Int J Not-for-Profit Law 7(4):3–27

Feitelson E (2013) The four eras of Israeli water policies. In: Becker N (ed) Water policy in Israel: context, issues and options. Springer, Dodrecht, Netherlands, pp 15–32

Feitelson E, Haddad M (eds) (2000) Management of shared groundwater resources: The Israeli-Palestinian Case with an international perspective. International Development Research Centre, Ottawa, and Kluwer Academic, Amsterdam

Feitelson E, Fischhendler I (2006) Politics and Institutions for groundwater management in a system-wide context. In: Presentation at the international symposium on groundwater sustainability, Alicante, Spain (January)

Feitelson E, Haddad M (1998a) Identification of joint management structures for shared aquifers: a cooperative Palestinian-Israeli effort. World Bank technical paper No. 415. World Bank, Washington, DC

Feitelson E, Haddad M (1998b) A stepwise open-ended approach to the identification of joint management structures for shared aquifers (Israel and Palestinian Authority). Water Int 23(4):227–237

Feitelson E, Tamimi A, Rosenthal G (2012) Climate change and security in the Israeli-Palestinian context. J Peace Research 49(1):241–257

Fischhendler I (2015) The securitization of water discourse: theoretical foundations, research gaps and objectives of the special issue. Int Environ Agreements: Polit Law Econ 15(3):245–255

Fleck J (2016) Water is for fighting over, and other myths about water in the West. Island Press, Washington, DC

Food and Agriculture Organization (FAO) (2008) Climate change: implications for agriculture in the Near East. Food and Agriculture Organization of the Middle East. In: Report for 29th FAO regional conference for the Near East (NERC/08/INF/5), Rome

Freimuth L, Bromberg G, Meyher M, Al-Khatib N (2007) Climate change: a new threat to Middle East security. Friends of the Earth Middle East, Amman, Bethlehem and Tel Aviv

Ge Y Li X, Cai X et al (2018) Converting UN sustainable development goals (SDGs) to decision-making objectives and implementation options at the river basin scale. Sustainability 10(4):1056. https://doi.org/10.3390/su10041056

Grover VI, Krantzberg G (2015) Transboundary water management: lessons from North America. Water Int 48(1):183–198

Haasnoot M, Middlekoop H, Offersmans A, van Beek E, van Deursen WPA (2012) Exploring pathways for sustainable water management in river deltas in a changing environment. Clim Change 115(3–4):795–819

Haddadin MJ (2002) Water in the Middle East peace process. Geogr J 168(4):324–340

Humpel D El-Naser H Irani K, Sitton J, Renshaw K, GleitsmanB (2012) A review of water policies in Jordan and recommendations for strategic priorities. USAID/Jordan/WRE

Inga C (ed) (2019) Climate change, water security, and national security for Jordan, Palestine, and Israel. EcoPeace Middle East, Amman, Ramallah, Tel Aviv

Jägerskog A (2003) Why states cooperate over shared water: the water negotiations in the Jordan River basin. Department of Water and Environmental Studies, University of Linköping, Linköping, Sweden

Karakaya M (2019) The failure of the NGO's for democratization: Egypt, Palestine & Jordan. Middle East Technical University, Graduate School of Social Sciences, Term Paper for Society in the Modern Middle East, Ankara, Turkey

Kay P, Mitchell B (2000) Measuring sustainability in Israel's water system. Water Int 25(4):617–623

Khalidi R, Daya W, Hashash AA, Jabareen A (2019) Political economy analysis of the Palestinian private sector. Palestine Economic Policy Research Institute (MAS), Ramallah

Kliot N, Shmueli D (1998) Real and ideal institutional frameworks for managing the common Arab-Israeli water resources. Water Int 23(4):216–226

Lautze J, Kirshen P (2009) Water allocation, climate change, and sustainable water use: the Palestinian position. Water Int 34(2):189–203

Lautze J, Reeves M, Vega R, Kirshen P (2005) Water allocation, climate change, and sustainable peace: the Israeli proposal. Water Int 30(2):197–209

Libiszewski S (1995) Water disputes in the Jordan basin region and their role in the resolution of the Arab-Israeli conflict. Center for Security Studies and Conflict Research, Swiss Federal Institute of Technology, Zurich, and Swiss Peace Foundation, Bern

Linton J (2010) What is water? The history of a modern abstraction. University of British Columbia Press, Vancouver

Lithwick H (2000) Evaluating water balances in Israel. In: Brooks DB, Mehment O (eds) Water balances in the eastern Mediterranean. IDRC Books, Ottawa, pp 29–58

Loehman E, Becker N (2006) Cooperation in a hydro-geologic commons: new institutions and pricing to achieve sustainability and security. Int J Water Resour Dev 2(4):603–614

Lonergan SC, Brooks DB (1994) Watershed: the role of fresh water in the Israeli-Palestinian conflict. International Development Research Centre, Ottawa

Medzini A, Wolf AT (2004) Towards a Middle East at peace: hidden issues in Arab-Israeli hydropolitics. Int J Water Resour Dev 20(2):193–204

Megdal SB, Varady RG, Eden S (eds) (2013) Shared borders, shared waters: Israeli-Palestinian and Colorado River basin water challenges. CRC Press/Balkema, Leiden, Netherlands

Nour R, Al-Saidi M (2018) Regulation reform process and perception in the Palestinian water sector. Water Int 43(7):908–925

Peralta RC, Luck AH, Hagan R (2011) Strategic optimization for implementing the Jordanian National Water Master Plan. Water Int 36(3):295–313

Rosenthal G, Katz D (2010) An economic analysis of policy options for water conservation in Israel. Kivun Consulting, Tel Aviv

Sa'ed RA (2010) A policy framework for trans-boundary waste-water Issues along the Green Line, The Israeli-Palestinian border. Int J Environ Stud 67(6):937–954

Salameh E (1990) Jordan's water resources: development and future prospects. J American-Arab Affairs 33:69–77

Shuval HI (1995) Towards resolving conflicts over water between Israel and its neighbours: the Israeli-Palestinian shared use of the Mountain Aquifer as a case study. Israel Aff 2(1):215–250

Shuval HI (2011) Comments on "Confronting water in the Israel-Palestinian peace agreement" by David Brooks and Julie Trottier. J Hydrology 397:146–148

Söderbaum P (2005) Actors, problem perceptions, strategies for sustainable development. Water policy in relation to paradigms, ideologies, and institutions. In: Biswas AK, Tortajada C (eds) Appraising sustainable development: water management and environmental challenges. Oxford University Press, New Delhi, pp 81–111

Söderbaum P (2009) Science, politics and water management for sustainability: economics as example. Water Int 34(4):432–440

Sowers J, Vengosh A, Weinthal E (2011) Climate change, water resources, and the politics of adaptation in the Middle East and North Africa. Clim Change 104(3–4):599–627

Staddon C, Scott CA (2018) Putting water security to work: addressing global challenges. Water Int 43(8):1017–1025

Subramanian A, Brown B, Wolf A (2012) Reaching across the waters: facing the risks of cooperation in international waters. In: Water papers. The World Bank, Washington, DC

Tal A (2006) Seeking sustainability: Israel's evolving water management strategy. Science 313(5790):1081–1084

Trottier J (1999) Hydropolitics in the West Bank and Gaza Strip. PASSIA Publications, Jerusalem

Trottier J (2007) A wall, water and power: The Israeli separation fence. Rev Int Stud 33(1):105–127

Trottier J (2008) Water crises: political construction or physical reality? Contemp Polit 14(2):197–214

Trottier J (2013) The social construction of water management at the intersection of international conflict: the case of Al Auja. Eurorient 44:161–181

Trottier J (2019) Palestinian water management—policies and pitfalls. PASSIA Bulletin, Jerusalem

Trottier J, Brooks DB (2013) Academic tribes and transboundary water management: water in the Israeli-Palestinian peace process. Sci Diplomacy 2:2

Trottier J, Rondier A, Perrier J (2019) Palestinians and donors playing with fire: 25 years of water projects in the West Bank. Int J Water Resour Dev. https://doi.org/10.1080/07900627.2019.1617679

Vickers A (2001) Handbook of water use and conservation. Waterplow Press, Amherst, MA, USA

World Bank (2009) Assessment of restrictions on Palestinian water sector development. Sector note for West Bank and Gaza, Washington, DC

World Bank Group (2018) Securing water for development in West Bank and Gaza: Sector Note. Washington, DC

Zeitoun, M (2007) The conflict versus cooperation paradox: fighting over or sharing of Palestinian-Israeli groundwater. Water Int 36(2):105–120

Zeitoun M (2009) The political economy of water demand management in Yemen and Jordan: a synthesis of findings. International Development Research Centre/WaDImena, Ottawa

Chapter 5
Supporting Palestinian Agriculture

Abstract All water management incorporates elements of both top-down and bottom-up authority and capacity. The latter have dominated Palestinian water management for many years. This chapter focuses on the extent to which the rural, agricultural portions of the country, which account for the great bulk of its water use, continue to use bottom-up structures despite the Oslo Agreements and despite Israel's military rulings. The current situation is further complicated by the arrival of agri-business interests that promote large date palm plantations that, on the one hand, provide Palestine with foreign exchange, and, on the other, pull water supplies and employment opportunities away from sharecropping farmers, thereby compromising livelihoods, housing security, and food security.

The current scientific discourse on Palestinian agriculture and water reflects the mainstream discourse concerning agriculture at the global scale. On one hand, reuse of waste water in irrigation is promoted as *creation of new water resources*, an innovation that is deemed useful to reduce abstraction from the aquifers. On the other hand, the notion of virtual water contributes to determining which crops are most suitable according to existing rainfall. Virtual water designates the quantity of water that is consumed when producing a crop (Allan 1998, 2003).[1] Within this logic, water-scarce countries should import water thirsty crops and should devote their scarce water resources to generating added value that will allow such importation (Hoekstra and Hung 2005). Given that logic, the transformation of the Jordan Valley into date palm cultivation (see Box 5.1) is perceived as a progress, which is exactly what will be shown to be both simplistic and inequitable in this chapter. Such cases are by no means limited to Palestine. Sara Wade (2018) demonstrates in her exploration of water justice in the context of the cut-flower industry in Kenya that it is all too easy for powerful commercial interests to co-opt senior water governance structures.

Improved policies for water in Palestine must accept two fundamental characteristics: First, Palestinian water has been managed locally, at village-scale, for thousands

---

[1] "Virtual Water:" A long-term solution for Middle Eastern countries? Available as: https://www.soas.ac.uk/water/publications/papers/file38347.pdf.

of years. Second, the rate of water use is much higher for agriculture than for households. Therefore, understanding Palestinian water management requires us to pay attention to the many ways that it developed in hundreds of villages, and also to pay greater attention to irrigation than to drinking water.

5.1 The Kerry Plan and Palestinian Agriculture

The Kerry Plan was elaborated as an Israeli-Palestinian peace plan under the Obama administration and included a large agricultural component within its economic plan. It considered Palestinian agriculture to be a "highly fragmented sector, with 111 K farm-holdings, of which 94%/are/smaller than 40 dunams" (Office of the Quartet Representative 2014). It proposed to promote integration through aggregation of small-holdings and vertical integration of producers and processors as well as international marketing. Aggregation into larger farms was seen as a way to enable farmers to begin farming as a business and to start planting higher value crops and intensifying farm management. The goal was to convert some 30,000 smallholdings into 600–1000 aggregated establishments, which the Initiative indicated would have to be export-oriented to be profitable.

The political aspect of the Kerry Plan collapsed, but its agricultural component seems to be still unfolding. This is unsurprising because the Kerry Plan gathered the tools deemed valid by the existing scientific discourse and policies to construct economic development in a future, independent state of Palestine. As far as agriculture is concerned, important components of this discourse and related policies were constructed between 2008 and 2012. They include priority given to the development of contract farming for export-oriented crops. The year 2008 was a crisis year globally that saw the price of foodstuff skyrocket, which fueled export-oriented contract farming around the world and which affected Palestinian development projects as well. Agricultural water projects were multiplied because irrigation stabilizes production and allows a farmer to meet calendars set by contract farming. The Kerry Plan integrated processes that were already promoted between 2008 and 2012 into a formalized economic development plan.

5.2 What Does "Supporting Palestinian Agriculture" Mean?

The current discourse on Palestinian agriculture was mostly developed by foreign diplomats or officials from the UN's Food and Agriculture Organization (FAO), and by scientists, whether Palestinian or foreign. Farmers themselves were rarely consulted. What does "supporting Palestinian agriculture" mean? Supporting whom? Does it mean supporting export-oriented Palestinian agribusinesses, some of which are already active in the Jordan Valley? Does it mean supporting smallholders who constitute the vast majority of farmers in the West Bank and the Gaza Strip? Does it mean helping existing farmers to thrive? If that is the case, does this mean land owners, many of whom live far from their land? Or farm owners who work their own land? Or is it sharecroppers, who commonly farm the same plots for many years without ever owning their own land? Or farmers who rent the land they cultivate? Or labourers, who may be year-long employees but most often are seasonal workers?

Raising such questions means considering land tenure and water tenure within an agricultural transformation. Tenure designates the relationship, either legally or customarily defined, between people, as individuals or groups, with respect to a resource (Hodgson 2016). It is a social construct that can be formal or informal, individual or collective. The mainstream discourse on Palestinian agriculture, whether produced by the scientific literature or by policy makers, has neglected land and water tenure when considering Palestinian agricultural development.

Land tenure in the eastern, mainly Palestinian, basin of the Mountain Aquifer is characterized by a high prevalence of sharecropping. Traditionally, open field cultivation within such a tenure meant that the sharecropper gave 50 percent of the revenue of the crop to the land owner. In the case of greenhouses, the sharecropper would give 75% of the revenue of the crop to the landowner (Trottier 2015). Thus, within sharecropping, the owner and the tenant both face the same risk of a poor crop or a collapse of the market price. This form of land tenure is rare in the western basin where smallholders cultivate their own land or rent land from owners who may reside in the village, a neighbouring town, or abroad.

Water tenure varies according to whether water is accessed through a well, a spring, or a wastewater treatment plant (WWTP). Traditionally, springs in the West Bank were systematically managed according to grass roots common property regimes, whereby right holders receive a water "turn"[2] every six to eight days for a set period of use. This way, the seasonal variability of the flow was shared in the same proportions among all users, whether they benefited from a small water turn or a large one. Abundance and scarcity were distributed in a similar manner among users of the spring. Wells that were dug before 1967 were also managed according to common property regimes through a *"sharikat el bir,"* a company established by local farmers who pooled their savings to drill a well in places where it could not simply be dug (Trottier 1999). Each farmer receiving such water, whether a member of the well company or not, got a monthly bill requesting him to pay a fee according to the number of hours of water he had used. A well company didn't aim to generate profit from selling the water. The fees covered the costs of operating the diesel pump and maintaining the well.

Such water tenure was never recognized by the 2002 Palestinian water law and the 2014 decree. Its Article 3 stipulates that all water is public property, and Article 31

---

[2]A water turn (as described in Trottier 1999) means that the full flow of a spring is directed to a farmer's land for a given amount of time, every 7 or 8 days (depending on the length of the water turn). One farmer could be entitled to the full flow of the spring from midnight to 10 min past midnight; then another farmer could be entitled to the full flow of the spring from ten after midnight until three in the morning. Obviously, the second farmer would have a larger share than the first farmer.

mentions a regulation should be prepared concerning prior use rights from springs or licensed quantity of water extracted from wells (Trottier and Perrier 2018). However, the law did not list or detail these prior use rights, nor has the regulation yet been prepared.

Since the Oslo Accord, water tenure has changed. The great number of unlicensed wells drilled in El Far'a Valley, for instance, dried up numerous springs (Trottier and Perrier 2018).[3] The common property regimes linked to these springs collapsed. The new unlicensed wells operate as private ventures, and the owner then commonly sells water to the people who formerly had traditional rights to the water of the very spring that his well contributed to drying up. In the Jordan Valley, one can easily get away with drilling and operating an unlicensed well in Area C, but not in Area A. However, elsewhere this pattern is often reversed. The advent of reuse of treated waste water now transforms water tenure further. As wastewater treatment plants have been built, the treated waste water is sold as irrigation water. WWTPs in West Nablus, Jericho, and Jenin were all running such reuse schemes in 2019. In all of these cases, water was brought to land that had never before been irrigated with ground water (Trottier et al. 2019b). Such reuse cannot possibly reduce abstraction from the aquifers.

The mainstream discourse on Palestinian agricultural development with its promotion of export-oriented contract farming, agribusiness, treated wastewater reuse, and date palm trees in the Jordan Valley did not factor in the specific forms of land and water tenure that existed in rural Palestine. As a result, it foresaw an economic improvement without assessing the disruptions such an agricultural transformation would bring to the livelihoods of those Palestinians involved in agriculture, especially smallholders and sharecroppers.

5.3   Water-Driven Palestinian Pioneer Fronts

Rural Palestine is hardly associated with expansion through agriculture. Yet interstitial "pioneer "fronts", or frontiers, exist in the West Bank among either uncultivated areas or areas used non-intensively that are nestled among villages, towns, Israeli settlements, and intensively cultivated areas (Trottier and Perrier 2018). Interstitial frontiers are transient phenomena whereby land that was previously either uncultivated or used in a non-intensive manner becomes suddenly the object of intensive agriculture. Either the pioneer front succeeds and an intensive form of agriculture becomes permanent, or it fails, in which case land use returns to its previous state.

The present interstitial agricultural frontiers in the West Bank are all driven by a newly accessed supply of water. In theory, three types of water supply-driven pioneer fronts may exist: surface, groundwater, and wastewater pioneer fronts. However, in the West Bank, all springs were developed for irrigation a long time ago, and therefore only groundwater and wastewater pioneer fronts currently exist. Groundwater pioneer fronts may rely on licensed wells or on unlicensed wells. In both cases, they

---

[3] Examples include Ein Far'a, Ein Mishke, and Ein Shibli.

are farmer led. However, waste water pioneer fronts are led by a network of state, donors, and investors.

Groundwater pioneer fronts that rely on pre-existing, licensed wells are mostly located in the western basin of the Mountain Aquifer. Such cases occur when urbanization or the separation wall pushes farmers further away from their village and from the existing licensed wells, higher in the mountain. The farmers then invest to build new terraces, pipes, and pumps to bring water uphill. Since 2011, they have often benefitted from donor support in building reservoirs. Drilling an unlicensed well is extremely difficult over the western aquifer. Therefore, these farmers must rely on a far-away, licensed well. The cost of pumping water uphill and the leaks along the way mean that the scientific discourse underlying the Kerry Plan's agricultural component categorizes such farmers as structurally inefficient.

Groundwater pioneer fronts relying on unlicensed wells are mostly located on the eastern and northern basins of the Mountain Aquifer. As opposed to licensed wells, unlicensed wells are private and have no written statutes at the time they are drilled. Some unlicensed well owners later manage to secure a license from the Palestinian Authority. The valley of El Far'a provides a prime example of such a transformation. The area of irrigated land has been multiplied by ten since the late 1990s. Rain-fed wheat and lentils have been replaced by irrigated vegetables. Well owners are aware that they are over-pumping the aquifer–very different from traditional management of the springs, which allowed sustainable use of water, as described by the Ottoman census and by foreign travelers in the 19th century (Guérin 1874; Hutteroth and Abdulfattah 1977). The present (*non*)-management of water in Al Far'a valley is the archetype of a *tragedy of the commons*.

Far too much importance has been given to a "cultural reluctance" against using wastewater. *Fatwas* by senior Moslem clerics permit properly treated waste water for use in irrigation (Abderraham 2000). In 2019, all of the treated waste water from the Jericho WWTP was used to irrigate Palestinian date palm trees. In Jenin, the full flow of the WWTP was used by farmers to cultivate fodder for their sheep, or to grow fruit trees (Trottier et al. 2019b). A waiting list exists of farmers ready to buy additional treated waste water once more houses will be connected to the WWTP in Jenin. In the Gaza Strip, an irrigation scheme was being built to pump untreated sewage infiltrated in the soil to reuse in fodder cultivation. Another scheme exists in West Nablus where treated waste water is used to irrigate trees. There is no cultural reluctance to use treated waste water provided that the level of treatment is adequate.

Box 5.1: Why Date Palm Plantations?

Date palms, whether Israeli or Palestinian, are hailed as the ideal crop for the Jordan Valley because they consume less than a third as much water as banana trees, which used to be the flagship cash crop in the area during the 1990s (Sonneveld et al. 2018). Medjoul dates fetch a high price on the international market, a price that is expected to remain inelastic even as supply increases.

Waste water, both from Palestinian households and from Israeli households, is bound to increase in the coming years. From 2011 to 2016, the PWA consistently purchased from Mekorot 60 percent of the water it distributed every year to Palestinian households. The quantity of water it purchased yearly increased by 16 million cubic meters within this period.

The most striking transformation of agriculture is taking place in the Jordan Valley where a tidal wave of date palm trees is sweeping over the land. Between 1999 and 2016, Palestinian date palm trees went from covering 25 ha to covering 1584 ha, and in the same time period Israeli (settler) date palm trees went from covering 524 ha to covering 2560 ha (Trottier et al. 2019a). While all Israeli trees were irrigated using treated waste water supplied by Mekorot from WWTPs and reservoirs along the Valley, Palestinian trees were irrigated using fresh water with the exception of those supplied by the Jericho WWTP. Palestinian agribusinesses rent land in the Jordan Valley, fence it, and replace sharecroppers with seasonal workers. This upheaval in land tenure means that the displaced sharecroppers can no longer live on the land they used to cultivate all year long. The seasonal labor for men lasts only two months of the year. Fencing the land also means that the poorest people can no longer pick nutritious weeds, such as *khubbezeh*, from irrigated fields, something that used to guarantee food security (Trottier and Perrier 2017).

As well, as mentioned above, at least a decade ago it was recognized in Israel that repeated irrigation with treated waste water leads to loss of soil structure (Travis et al. 2010; Tal 2016). Desalination of treated waste water is possible, but of course adds to costs. Most Palestinian date palm trees are not irrigated with waste water so that is not currently a problem. However, as salt deposits are never washed off the land, the soil around the tree becomes increasingly saline. This is not a problem for agribusinesses that rent the land and plan to move to another plot after 40 years, but it is a long-term problem for protection of the environment and for future agricultural productivity.

In summary, the West Bank is a small place: 5655 km^2. The great number of water projects carried out since 1994 in such a small area has massively impacted spatial, institutional, and sectoral water trajectories. As a result, the overall impact of all donor-supported water projects has been greater than the sum of the individual projects. Understanding how these projects have altered the many trajectories of water allows understanding of how they have altered the way Palestinian society structures its interactions with water. Perhaps most notable is the way that this transformation of Palestinian agriculture, which is hailed as a move toward more efficient water use, has severely compromised the livelihoods of former sharecroppers at the same time as it enriches new well owners and agribusinesses.

Further, the claim that date palm trees "free up" water is questionable. By 2016, 786 ha of Palestinian date palm trees were planted on previously irrigated land. However, 778 ha of date palm trees had been planted on previously unirrigated land (Trottier et al. 2019a). As the volume of date plantations keeps growing, the switch to a crop with low virtual water content has clearly not "freed up" water for other crops.

Finally, almost all irrigation in the Jordan Valley occurs through drip systems, which is generally a positive action but has some downsides. Traditional irrigation is mainly labour intensive, which is fine for sharecroppers, whereas drip irrigation is mainly capital intensive and therefore better suited to owner-operated farms. As well, experience elsewhere shows that potential quantitative water savings are often lost as farmers shift to intensive cultivation of more valuable crops or extend cultivation to previously rain-fed fields (Birkenholtz 2017). Trottier and her colleagues show how the same thing occurs in the West Bank (Trottier and Perrier 2017, 2018; Trottier et al. 2019b). In poor areas, it may direct more water into one trajectory at the expense of another trajectory. Often that new trajectory leads the water to a monetized use, whereas the former trajectory brought water to uses that are not monetized but essential for local food security.

The switch to date palm trees has certainly generated foreign currency in Palestine, but it has simultaneously disrupted the economic livelihoods of many residents of the Jordan Valley and impaired their previous environmental sustainability.

References

Abderraham WA (2000) Water demand management and Islamic water management principles: a case study. Int J Water Resour Dev 16(4):465–473

Allan JA (1998) Virtual water: a strategic resource, global solutions to regional deficits. Groundwater 36(4):545–546

Allan JA (2003) Virtual water—the water, food, and trade nexus: Useful concept or misleading metaphor? Water Int 28(1):106–113

Birkenholtz T (2017) Assessing India's drip-irrigation boom: efficiency, climate change and groundwater policy. Water Int 42(6):663–677

Guérin V (1874) Description géographique, historique et archéologique de la Palestine accompagnée de cartes détaillées. Oriental Press in 1969, Paris in 1974, Amsterdam

Hodgson S (2016) Exploring the concept of tenure. FAO, Land and Water Discussion Paper 10, Rome

Hoekstra AY, Hung PQ (2005) Globalisation of water resources: International virtual water flows in relation to crop trade. Glob Environ Change 15(1):45–56

Hütteroth W-D, Abdulfattah K (1977) Historical geography of Palestine, Transjordan and Southern Syria in the late 16th century. Geographische Arbeiten, Erlangen

Office of the Quartet Representative (2014) Initiative for the Palestinian economy: agriculture. Jerusalem. http://www.quartetoffice.org/files/server/agriculture.pdf

Sonneveld BGJS, Marei A, Merbis MD, Alfarra A (2018) The future of date palm cultivation in the Lower Jordan valley of the West Bank. Appl Water Sci 8:113

Tal A (2016) Rethinking the sustainability of Israel's irrigation practices in the drylands. Water Res 90(1):387–394

Travis M, Wiel-Shafran A, Weisbrod N, Adar E, Gross A (2010) Greywater reuse for irrigation: effect on soil properties. Sci Total Environ 408(12):2501–2508

Trottier J (1999) Hydropolitics in the West Bank and Gaza Strip. PASSIA Publications, Jerusalem

Trottier J (2015) Le rapport à l'eau et à la terre dans la construction de territoires multisitués: le cas palestinien. Espace Géographique 44(2):103–114

Trottier J, Perrier J (2017) Challenging the coproduction of virtual water and Palestinian agriculture. Geoforum 87:85–94

Trottier J, Perrier J (2018) Water-driven Palestinian agricultural frontiers: the global ramifications of transforming local irrigation. J Polit Ecol 25(1):1–21

Trottier J, Leblond N, Garb Y (2019a) The political role of date palm trees in the Jordan Valley: The transformation of Palestinian land and water tenure in agriculture made invisible by epistemic violence. Nat Space, Environment and Planning E forthcoming

Trottier J, Rondier A, Perrier J (2019b) Palestinians and donors playing with fire: 25 years of water projects in the West Bank. Int J Water Resour Dev. https://doi.org/10.1080/07900627.2019.1617679

Wade S (2018) Is Water security just? Concepts, tools and missing links. Water Int 43(8):1026–1039

Chapter 6
Supplementary Approaches to Shared Transboundary Water Management

The fact that there has been widespread international cooperation over water should not allow policy-makers to underestimate the complexity of the relationship between water and national security.
Axworthy and Sandford (2012, p. 12)

Abstract The EcoPeace Proposal is considered to be the most attractive way to move Israel, Palestine, and Jordan toward a final status agreement for their shared water resources. However, they are not the only way, and this chapter focuses on four supplemental or alternative ways for dealing with regional water issues: the Mountain Aquifer study shows how the EcoPeace Proposal could be tried first at smaller scale; the Lower Jordan Basin project is an actual example of what can be achieved in a transboundary region that is highly sensitive ecologically but with a large economic potential; the Red Sea-Dead Sea Water Conveyance (RSDSWC) project is a proposed way of meeting both water and electricity shortages, mainly in Jordan, by conventional means but at massive cost; finally the Water-Energy Nexus is both a modern response to water insecurity and an opportunity for regional integration. The Chapter concludes with a note on the experience of EcoPeace in its presentations before the United Nations Security Council.

6.1 Mountain Aquifer Authority

It has been said that the Mountain Aquifer was designed by an evil water god. Geologically, it is a karstic (limestone) aquifer that is complexly fractured and permits relatively fast flow of water through it. Politically, it underlies the pre-1967 (and likely post-peace agreement) boundary between Israel and Palestine (see Fig. 2.1). About 90% of the catchment area for the aquifer lies on the Palestinian side of the boundary, but two of its three sub-basins flow naturally to the Israeli side. The result is an aquifer that would be a political problem if it underlay the boundary between the United States and Canada.

The concept for an Israeli-Palestinian study of the Mountain Aquifer was initially discussed at The 1st Israeli-Palestinian International Academic Conference on Water, which took place at the Swiss Federal Institute of Technology in Zurich, Switzerland, on 10–13 December 1992.[1] The study was subsequently funded by Canada's International Development Research Centre and was assisted in Israel by the CRB Foundation (Canada does not provide economic assistance to Israel).

In the hydrosocial setting of the Mountain Aquifer, key issues are more institutional than technical. Even before the Oslo peace talks, analysts from both nations were beginning to investigate options for joint management of the Mountain Aquifer, something for which there is only limited experience elsewhere in the world and for which policy and legislation are only now being developed (Puri and Aureli 2005). Once the formal peace process started, this research was construed as "academic" or "2nd track" activity complementing political bilateral and technical multilateral tracks. Some of the people who participated in diplomatic negotiations also participated as analysts in the Mountain Aquifer study.

Almost from the start of the research, the research team defined four basic goals of joint management:

1. Resource Protection—to avoid loss of water quality;
2. Crisis Management—to respond to both spills and drought;
3. Economic Efficiency—to approach the results that would come with a private market;
4. Integrated aquifer management—to cover social, physical and environmental aspects, possibly with regulatory powers.

No attempt was made to create a blueprint. Just as the research was an example of learning by doing, so also would be the management process. In many ways, this approach reflected the Dublin Principles, which were formulated in 1992 and which attempted to find a way to avoid the unfortunate dichotomy between water as an economic good and water as a human right.[2]

---

[1] *Note by DBB*: As someone who attended the 1st conference and presented one of two keynote talks, I can indicate how sensitive it was by recalling that approval for the Conference required intense negotiations and was only received from the Israeli government and the Palestinian Liberation Organization on the day before the Conference was supposed to begin. It was a dozen years before the 2nd Israeli-Palestinian Conference was held in Antalya, Turkey, from 10–14 October 2004, which I also attended. It is significant that the word "academic" had been dropped—it was no longer needed to assure approval—and that, despite the title, its scope was enlarged from Israel and Palestine to the whole Middle East. Results of the 1st Conference were reported in Issac and Shuval (1994), and abstracts from the 2nd Conference by IPCRI (2004).

[2] The Dublin Statement on water and Sustainable Development was agreed at an International Conference on Water and the Environment in January 1992, a preparatory meeting of the United Nations Conference on Environment and Development that was held later that year. The Dublin Statement included the following four principles:

1. Fresh water is a finite and vulnerable resource, essential to sustain life, development and the environment.
2. Water development and management should be based on a participatory approach, involving users, planners and policy-makers at all levels.

As the study progressed, the teams found that often the most controversial issues turned out to be sectoral, not national. If Israeli farmers suffer, so too will Palestinian farmers. It also proved easier to respond to quality issues rather than quantity allocation. Indeed, quality issues commonly stimulated cooperation over quantity allocation. It was also more productive to focus on the process of creating a management structure rather than on specific final goals.

The study went through four phases, and has now been completed. Its output appears in a number of reports, each published jointly by The Palestine Consultancy Group in East Jerusalem and The Harry S. Truman Institute for the Advancement of Peace at The Hebrew University of Jerusalem. In addition, a book edited by Feitelson and Haddad (2000) was published, as were summaries of the methods and results by the same authors (1998a, b).

6.2 Lower Jordan River Basin Protection and Development

The Jordan River originates mainly from a group of springs in Israel about five kilometers south of the Lebanese border, as well as from smaller springs in Lebanon and on the Golan Heights in Syria (though partially occupied by Israel). It then flows southward as the Upper Jordan into the Sea of Galilee (Lake Kinneret), and below the lake continues southward as the Lower Jordan until it empties into the Dead Sea. Israel, Palestine and Jordan are all riparian to the Lower Jordan River, and its basin of about 18,000 km^2 has been historically seen as a significant religious, economic, and environmental site with unique importance to the three Abrahamic religions. Syria and Lebanon are also riparians, but they do not participate in current discussions about the future of the basin.

In the past half century or so, international, national and local actors have competed with proposals and activities for the region's development. As a result, in perhaps the single most comprehensive study available prior to 2010, Van Aken et al. (2009, p. 42) declared the basin to be "closed"[3] "as most of the water is mobilized and depleted." Diversion of the Jordan's water by Israel, Syria and Jordan for domestic and agricultural uses has decreased the outflow into the Dead Sea from 1200–1300 million cubic metres (MCM) per year before 1950 to only 70–100 MCM per year today. During the same period, the river has lost more than half its biodiversity,

---

3. Women play a central part in the provision, management and safeguarding of water.
4. Water has an economic value in all its competing uses and should be recognized as an economic good.

   Further information on the Dublin Statement is available at:
   https://en.wikipedia.org/wiki/Dublin Statement.

[3]Hydrologically, a basin is defined as closed when the inflow has no outlet and the water only escapes the basin by evaporation upward or percolation downward. That is not the meaning used here. In socio-economic terms a basin is closed when there is little or no water that can be withdrawn for any use. That is the case for the Lower Jordan River basin.

mainly as a result of fewer fast flows and floods as well as increased salinity (Gafny et al. 2010).

The Lower Jordan River Basin project began from recognition that the river could not be restored to ecological health unless some currently diverted fresh water was restored, with follow-on economic benefits from religious visits and tourism (Gafny et al. 2010; Baltutis 2011; Hylton et al. 2012; Safier et al. 2011). Several major governing bodies were given roles in the rehabilitation of the Basin. In Israel, the Lower Jordan River Drainage Authority was created to promote the conservation, rehabilitation, and development of the river area. In Jordan, the Jordan Valley Authority and the Water Authority of Jordan are responsible for water resources development. Palestine does not have direct control over the river itself as a result of Israeli military controls on Palestinian land bordering the river, but they obviously do have interests that it vigorously expresses.

Somewhat before the start of the Lower Jordan River project, Germany had financed the GLOWA Jordan River study. GLOWA's website[4] describes the project as "an interdisciplinary and international research project providing scientific support for sustainable water management in the Jordan River region." The project depended upon classical mathematical modelling instead of field observations, which lessened its potential value. Nevertheless, after it closed in early 2015, it left an archive of information and data, some of which has proven useful for those wanting to pursue protection and development further.

Despite initial steps to protect and develop the Jordan River basin, the persistence of damage necessitated further interventions. Perhaps most significant is the Jordan Valley Master Plan for Sustainable Development (2015), which was prepared by EcoPeace Middle East. It analyzes existing national plans in Jordan, Palestine, and Israel and proposes a picture of the basin in terms of land use, different economic sectors, available resources, population, and current governance for the years 2015, 2025 and 2050. Though by no means fully accepted by any of the government agencies, the Master Plan does provide a viable route forward toward an ultimate goal of regional integration, sustainable water use, and political peace for the Lower Jordan River Basin.

6.3   Red Sea-Dead Sea Water Conveyance and Desalination Project

The notion of a canal or tunnel from the Mediterranean or Red Sea to the Dead Sea has a long and distinguished (if somewhat fruitless) history.

(rlcohen@gmail.com; 23 June 2008; Building the wrong canal)

---

[4]www.glowa.jordan.river.com; GLOWA is a German acronym meaning that, in English, means Global Change in the Hydrological Cycle.

The Dead Sea region is internationally known for its unique geographical, bio-logical, and historical values. It is the lowest point on the surface of the earth and the world's saltiest deep water body. The Dead Sea waters, rich in a wide variety of minerals, are famous for their therapeutic qualities and the wetlands surrounding the Sea are blessed with unique flora and fauna. However, the Dead Sea is drying up at an alarming rate. The primary cause of its rapid decline is diversion of water that used to pour into it from the Jordan River (Gafny et al. 2010). Next in importance is the mineral extraction by the Dead Sea Works in Israel and the Arab Potash Company in Jordan. They operate industrial-scale solar evaporation ponds that, according to Israel's Ministry of Environment,[5] are responsible for about one-third of the deple-tion of Dead Sea waters. Since the 1960s, the level of the surface of the Dead Sea has fallen over 25 meters and is continuing to drop by over 1 m per year. One result is the slowly drying shore with over 5000 sinkholes, most of them on the flat western Israeli side.

In 2002 at the World Summit for Sustainable Development held in Johannesburg, South Africa, the governments of Israel and Jordan advanced the idea of building a water link from the Red Sea to the Dead Sea. The draft proposal, which came to be known as the Red Sea-Dead Sea Water Conveyance (RSDSWC), involved the transfer of 2 billion cubic meters of water from Aqaba to the Dead Sea, plus construction of a desalination plant near the Dead Sea to produce about 800 MCM of potable water, mostly destined for household water needs in Amman, Jordan's capital, and to release over 1 billion cubic meters of brine into the Dead Sea to raise its surface level to previous heights. RSDSWC was a top-down proposal and its tripartite support—Palestine began to play an active role after the meeting at the White House in 2013–concealed the fact that each country argued its case from a different narrative and ignored any shift toward bottom-up arguments (Hussein 2017).

Early estimates of project cost started at $4 billion, and over time they have grown. No wonder that the project was described as a "mega water project for water in the Middle East at a mega cost" (Hersh 2005, cited by Spiritos and Lipchin, p. 120). A similar project had been rejected by the World Bank in the 1990s on the basis that neither the Israeli nor the Jordanian economy would be able to sustain such expensive infrastructure. Indeed, every nut and bolt would need to be replaced within a few decades because it was carrying sea water, which is much more corrosive than fresh water.

Concerned that the World Bank feasibility study looked only at alternative designs *for* the RSDSWC, but not at alternatives *to* it, a number of NGOs, among them EcoPeace, argued for a second feasibility study that would focus on other means of achieving the twin goals of drinking water to Amman and stabilization of the Dead

---

[5]Israeli Ministry of Environment, Policy Paper on the Future of the Dead Sea; 2005, Israeli Knesset Protocol on Minister's question regarding water usage by Dead Sea Works Mineral Industry; 2013.

Sea.[6] The second study indicated that there were indeed alternatives available with less environmental risk and at lower costs. They appeared in a large final report—the Executive Summary itself is 62 pages long—that avoids any single choice, but rather shows how each of the 20 or so alternatives or combinations of alternatives compares against a variety of goals.

Discussions about the RSDSWC continued until December 2013 when a Memorandum of Understanding (MoU) for a pilot "Red Sea Dead Sea Water Project" was signed in Washington by the three parties. This new project, very different from the mega-project, envisions a water exchange between Jordan and Israel, whereby Jordan would desalinate 80 MCM near Aqaba and sell some 50 MCM annually of this desalinated water to Eilat in the south of Israel; and in exchange Israel would sell 50 MCM water annually from the Sea of Galilee to Irbid and Amman in the north of Jordan (in addition to the 50 MCM that Israel is currently sending to Jordan as its share of winter flows into Lake Tiberius). As part of the MoU, Israel also agreed to sell 33 MCM of water to the Palestinians.

Even with the MoU, questions remain about mixing of Red Sea brine with the unique mineral waters of the Dead Sea.[7] The World Bank study claims that the maximum amount of brine that the Dead Sea could receive without causing damage is 400 MCM. Based on this assessment Jordan and Israel pledged not to exceed a release of 300 MCM, but that does not determine which of several alternatives for brine disposal would be adopted. The World Bank study of alternatives does conclude by restating that any final proposal must incorporate two complementary components: partial rehabilitation of the Jordan River's flow into the Dead Sea, and changes in mineral industry practices. Both are considered essential if the Dead Sea is to be saved.[8]

6.4   Water–Energy Nexus Among Israel, Jordan, and Palestine

It is expected that by 2030 Israel, Jordan, and Palestine will each have to face a gap between water needs and available water deliveries that is so large that the region will not be able to live solely on natural fresh water resources. Both the Jordan-Israel Peace Treaty of 1994 and the Oslo II Accord of 1995 contain language on the need for cooperation for development of new sources of water.

---

[6]COYNE-ET BELLIER in association with TRACTEBEL ENGINEERING and KEMA, RED SEA—DEAD SEA WATER CONVEYANCE STUDY PROGRAM FEASIBILITY STUDY Draft Final Feasibility Study Report Summary, July 2012.

[7]Gavrieli Itay, Lensky Nadav G., Dvorkin Yona, Lyakhovsky Vladimir, and Gertman Isac, A Multi-Component Chemistry-Based Model for the Dead Sea: Modifications to the 1D Princeton Oceanographic Model, Ministry of National Infrastructures Geological Survey of Israel, EcoPeace Middle East, USAID, Report GSI/24/2006.

[8]https://www.nationalgeographic.com/environment/2018/12/israel-jordan-dying-dead-sea-pollution-tourism/.

Desalination has become the primary source of nonconventional water for Israel. However, current processes for desalination remain based on fossil fuels and are therefore major sources of greenhouse gases (GHGs). Israel, Jordan and Palestine have committed to reducing GHGs, and therefore the region's great opportunity lies in its large areas of mainly unpopulated lands that are not suitable for agriculture but very suitable for solar energy facilities. Such resources are found only in Jordan.

The Water-Energy Nexus (WEN) is both an institutional and infrastructural solution, which foresees a water and energy community among Jordan, Palestine and Israel. Jordan would use its areas of open land to harness solar energy, which could then supply the renewable energy needed to desalinate sea water in coastal areas. A prefeasibility study (Katz and Shafran 2017), funded by EcoPeace Middle East and Konrad-Adenauer-Stiftung, together with a subsequent summary (Katz and Shafran 2019) determined that the exchanges are technically feasible with potential benefits for each of the countries, at a cost that is both significantly lower and responsible for much less GHG emission than any other alternative. These results are so striking that it is worth quoting this excerpt from the full study (p. 6):

> In 2030, with expected population of nearly 30 million people, the region will need an additional 4 million cubic meters (MCM) of water annually just to maintain current levels of domestic consumption. The cost of providing this water in coastal areas of Palestine and Israel serving 50% and 70% of population respectively could be provided at a cost of roughly US$0.65 per cubic meter ($m^3$), while the cost of providing water to urban centers in Jordan such as Irbid and Amman serving 80% of population, would range from between US$0.93–1.18/$m^3$. This estimate suggests that WEN would provide the cheapest marginal cost of water currently available to Jordan.

> The study also shows that supplying 20% of the region's projected energy demand in 2030 with solar energy could be accomplished at US$0.05–0.07 per kilowatt hour, a cost that is cheaper than the most efficient current fossil fuel production, even without considering the environmental costs of burning fossil fuels.... While Palestine and Israel have limited available open spaces for such projects, Jordan has plenty, and production at this scale would require only 0.1% of total Jordanian land area.

Of course, there will be some downsides. Notably, building and operating industrial size solar facilities in the middle of the desert is going to harm the local ecology, including animals and plants that have existed for eons largely free from human interference. Some analysts are clearly fearful of their rapid and widespread deployment in desert areas. The first few sentences of the abstract of an article by Moore-O'Leary et al. (2017, p. 385) are certainly disturbing:

> Renewable energy development is an arena where ecological, political, and socioeconomic values collide. Advances in renewable energy will incur steep environmental costs to landscapes in which facilities are constructed and operated. Scientists—including those from academia, industry, and government agencies—have only recently begun to quantify trade-offs in this arena, often using ground-mounted, utility-scale solar energy facilities (USSE ≥ 1 MW) as a model.

Less analytical, but in a way more persuasive and certainly more beautiful is the article by Alagona and Smith (2012), which comes with photographs by Christopher Woodcock.

The EcoPeace/KAS study cited above (Katz and Shafran 2017) takes it as given that Jordan and Palestine are going to have to follow Israel's experience and promote desalination just to maintain current levels of lifestyles and livelihoods for their growing populations. They then compare environmental effects for fueling the desal plants with natural gas versus using solar energy, and show that environmental protection argues decidedly in favour of the latter. This comparison is not naïve. Among other things they take account of some of the difficulties of dealing with solar photovoltaics, including the need to wash them to remove dust and sand, and the facts that panels generally have no more than 20 year life spans and that their production rates decline over time.

Other analysts question whether the energy return on investment (EROI)—that is, the energy produced over the life time of the solar cell compared with the energy required for its production–is as positive as commonly assumed. The same sort of question is also asked as to whether GHG emissions from solar cells over their lifetimes really are dramatically less than for fossil fuels. The answers to these questions depend on technologies for producing the cells, which are changing rapidly. For example, modern photovoltaic cells do not need silicon as pure as advanced microprocessors, and they can use thinner slices of silicon bars, either of which will both improve their EROI and reduce their GHG emissions per produced kWh. On the other hand, if the original source for electricity is coal, those improvements would at least be moderated if not eliminated.[9]

Though Jordan may share some ecological characteristics with deserts in Southwestern United States, it does not share the population density nor any expectation of a large demand for desert living in gated communities by well-off Americans. With adequate ecological assessment and appropriate planning, and the opportunity to leave more than 99% of Jordan free of solar collectors, it should be possible to avoid sensitive ecological areas and to protect endangered species. Those measures will add to costs, but they should not preclude going forward with a few WEN operations, and, as a condition for any license, to insist upon ongoing measurement of ecological impact both during the construction phase and during the operational phase. EcoPeace Middle East's Jordan office has been promoting the concept, notably with a workshop in July 2019 that brought together climate scientists, hydrologists, NGOs, and university professors, in an attempt to better understand how to communicate scientific ideas and plans about WEN and other options on water security to government officials.

By fostering an interdependent geo-political relationship, each partner in a WEN project would be able to use its comparative advantages to provide a more sustainable option to meet water scarcity than could any of the three countries working alone. This functionalist model of regional integration has been proven successful by the European Coal and Steel Community, which was established shortly after World War 2 and which set a precedent for regional cooperation over shared resources.

---

[9]We must thank Dr Jean Verdier, who serves on the Board of the Association Française de l'Eau, de l'Irrigation et du Drainage, for his assistance with information and citations offering comparisons of solar photovoltaics with conventional sources of electricity.

If similarly successful for water and energy for Israel, Jordan and Palestine, WEN could be extended to other countries in MENA.

At present, solar facilities are being built across the region, most actively in United Arab Emirates, Saudi Arabia, Morocco, and Egypt,[10] but most are renewable alternatives for fossil-fuel generated electricity. Though commendable for reducing GHG emissions, none offers the potential of WEN, something that was already recognized by *The Economist* in an article about WEN-based utilities in the Middle East with the subtitle, "An environmental proposal with political overtones."[11]

Jordan has begun making a start at appropriate renewable energy projects. With no significant sources of fossil fuels, wind and sun are its only reliable sources of low-cost electricity. The link with water is obvious once one notes that over 14 percent of the nation's total power production is given to energy projects. The nation has established a goal of producing 20 percent of the water sector's electricity requirements renewably by 2021.[12]

6.5 EcoPeace at the UN Security Council

To conclude Chapter 6 on a particularly positive note, EcoPeace's work came to the attention of the United States' representative at the 8517th United Nations Security Council session on the Middle East on 29 April 2019. This was the session where two EcoPeace co-directors—Gidon Bromberg for Israel and Nada Majdalani for Palestine—were invited to deliver a talk on environmental peacebuilding and where they called on nations to shift resource management from conflict to cooperation across borders and highlighted the impact this could have on regional security. Mr. Rodney Hunter, Political Coordinator for the United States Mission to the United Nations, spoke after their talk, and commented that, "It is indeed heartening to hear that Israelis, Palestinians, and Jordanians are working together on ways to address shared challenges in energy and water."[13] Mr Hunter was speaking specifically about WEN, but by implication he was referring to the benefits of all forms of cooperation among the three countries.

---

[10]https://www.ynetnews.com/articles/0,7340,L-5448709,00.html.

[11]Print edition, Issue of 16 January 2016.

[12]http://www.jordantimes.com/news/local/20-cent-water-sector%E2%80%99s-electricity-soon-come-renewable-sources.

[13]https://usun.state.gov/remarks/9042.

References

Alagona PS, Smith CF (2012) Mirage in the making: Mojave dreams. Boom J Calif 2(3):25-44; Photographs by Christopher Woodcock

Axworthy T, Sanford R (2012) The global water crisis: framing the issue. In: Biggs H (ed) The global water crisis: addressing an urgent security issue. Papers for the InterAction Council: UNU-INWEH,. Hamilton, ON, Canada, pp 2–7

Baltutis J (2011) Economic benefits of access to a healthy Lower Jordan River for the Palestinian Economy. Friends of the Earth Middle East, Amman, Bethlehem and Tel Aviv. http://hispagua.cedex.es/sites/default/files/hispagua_documento/documentacion/documentos/jordan_saludable.pdf

Feitelson E, Haddad M (1998a) Identification of joint management structures for shared aquifers: a cooperative Palestinian-Israeli effort. World Bank technical paper no. 415. World Bank, Washington, DC

Feitelson E, Haddad M (1998b) A stepwise open-ended approach to the identification of joint management structures for shared aquifers (Israel and Palestinian Authority). Water Int 23(4):227–237

Feitelson E, Haddad M (eds) (2000) Management of shared groundwater resources: The Israeli-Palestinian case with an international perspective. International Development Research Centre, Ottawa, and Kluwer Academic, Amsterdam

Gafny S, Talozi S, Al Sheikh B, Ya'ari E (2010) Towards a living Jordan River: an environmental flows report on the rehabilitation of the Lower Jordan River. Friends of the Earth Middle East, Amman, Bethlehem, Tel Aviv. http://foeme.org/uploads/publications_publ117_1.pdf

Hylton E et al (2012) Take me over the Jordan: Concept document to rehabilitate, promote prosperity, and help bring peace to the Lower Jordan valley. Friends of the Earth Middle East, Amman, Bethlehem, Tel Aviv

Hussein H (2017) Politics of the Dead Sea canal: a historical review of the evolving discourses, interests, and plans. Water Int 42(5):527–542

IPCRI (Israel/Palestine Center for Research and Information) (2004) In: Abstracts, vol 2nd Israeli-Palestinian international academic conference on water for life in the Middle East. Jerusalem

Issac J, Shuval HI (eds) (1994) Water and peace in the Middle East: In: Proceedings of the first Israeli-Palestinian international academic conference on water. Elsevier; Studies in Environmental Science No. 58, Amsterdam

Katz D, Shafran A (eds) (2017) Water-energy nexus. A pre-feasibility study for Mid-East water-renewable energy exchanges. EcoPeace Middle East, Amman, Ramallah, Tel Aviv, and Konrad-Adenauer-Stiftung, Amman

Katz D, Shafran A (2019) Transboundary exchanges of renewable energy and desalinated water in the Middle East. Energies 12(8):1455

Moore-O'Leary KA, Hernandez RR, Johnston DS et al (2017) Sustainability of utility-scale solar energy—critical ecological concepts. Front Ecol Environ 15(7):385–394

Puri S, Aureli A (2005) Transboundary aquifers: a global program to assess, evaluate, and develop policy. Ground Water 43(5):661–668

Safier G, Arbel Y, Bromberg G, Ya'ari E (2011) Road map for the restoration of the Lower Jordan River. Friends of the Earth Middle East, Amman, Bethlehem, Tel Aviv. http://foeme.org/uploads/DHV_Report_Ex_Summary_E_11_2011%281%29.pdf

Spiritos E, Lipchin C (2013) Desalination in Israel. In: Becker N (ed) Water policy in Israel: context, issues and options. Springer, Dodrecht, Netherlands, pp 101–123

Van Aken M, Molle F, Venot J-P (2009) Squeezed dry: the historical trajectory of the Lower Jordan River Basin. In: Molle F, Wester P (eds) River basin trajectories: societies, environments and development. CAB International, Oxfordshire, UK, pp 20–46

Chapter 7
Moving Water from Last to First in the Peace Process

> *The increasingly urgent reform of water allocation is challenged by the complexity of the political dimension, in particular the need to reconcile often competing objectives such as food and energy security and green growth.*
> Hellegers and Leflaive (2015, p. 273).

Abstract This chapter summarizes the status of water negotiations among the three nations of Israel, Palestine, and Jordan as they were to now and as they could be with a broader interpretation and understanding of the role of water in human society. They all have much to gain from reaching a formal, equitable and sustainable agreement, but even more to lose from continued conflict and disruption. Indeed, they could become a model for transboundary water management among (formerly hostile) nations elsewhere in the world.

Since the start of the Oslo process in 1993, solving the water issue has been held hostage to lack of progress on other core issues of the peace process. This stalemate is as remarkable as it is sad, given that for a long time now most analysts agree that water issues are solvable and will result in the Palestinians receiving a larger proportion of shared Israeli-Palestinian water (Hadi 2003; Shuval 2007; Shuval and Dweik 2007).

7.1 From Then to Now

If resolution of water issues was possible more than a decade ago, it is all that more so today when large-scale desalination has shown itself capable of providing drinking water for a large share of household uses in Israel, and doing so at reasonable prices. Certainly, resolving water issues at this time should be less contentious than doing so as part of a final status agreement that must also deal with borders, refugees, Israeli settlements, and the status of Jerusalem.

Evidence exists at local, national and international levels in support of the demand for action now on water. At the local level, EcoPeace's 25 years of experience with water cooperation between communities on opposite sides of a border has yielded notable examples of confidence building. Its Good Water Neighbors Project demonstrates further that the more frequent and intensive the cooperation, the greater the mutual understanding—and the greater the understanding, the more acceptable the results (Sagive et al. 2012).

At the national level, one of the notable changes in many years has been the independent decision of the Israeli Water Authority to commit to releasing 30 Million Cubic Metres (MCM) per year into the River. This volume will make only a modest improvement as analysis indicates that 400 to 600 MCM are needed for full rehabilitation of the Jordan River (Gafny et al. 2010). However, a longer term goal is to release 200 MCM of desalinated water by 2022, about half way to the minimum needed for ecological restoration.

In parallel, strenuous efforts on both the supply and the demand sides allowed the Jordanian Ministry of Water and Irrigation to announce a reduction in the nation's water deficit from 405 MCM in 2017 to 373 MCM in 2018, which is remarkable given the 1.4 million Syrian refugees currently living in Jordan.[1] Moreover, with growing recognition that longer and more frequent droughts are the new normal, all Middle East and North African countries have been urged strongly to adopt the United Nation's Food and Agriculture Organization's call to shift drought management policy from short-term emergency response to long-term reduction of risks and greater resilience (Bazza et al. 2018).

A major opportunity to increase Israeli water sales to Palestine was lost in early 2019 when the new United States Anti-Terrorism Clarification Act (ATCA), which had been passed in 2018, came into force. About a year earlier, Donald Trump's envoy to the Mideast, Jason Greenblatt, had brokered a water-sharing agreement between Israel and the Palestinian Authority. However, ATCA empowers Americans to sue foreign aid recipients in U.S. courts over alleged complicity in "acts of war." As one presumably unintended result of the law, the Palestinian Authority announced it would stop taking foreign aid from the United States, which in turn caused the US Agency for International Development to stop operating in the West Bank and the Gaza Strip in February, and that was the end of the Greenblatt agreement. Wildeman and Tartir (2019) responded that, so far from being a problem, the end of US aid to Palestine would be beneficial to Palestine and to a future peace agreement.

At the International level Aaron Wolf (1999a, b, 2000) reports that most international negotiations over water during the past century have proceeded on the basis of each side recognizing the "needs" of the other side(s), rather than disputing *a priori* principles or rights. Megdal et al. (2013) expand the issue of needs to one of the most important conclusions for dealing with shared water (p. 275):

---

[1] http://www.jordantimes.com/news/local/water-deficit-drops-8-cent-thanks-more-efficient-supply.

It is essential to find and implement solutions that meet the needs of neighboring societies.. ..
In this context, it becomes important to ensure that all parties, including some not traditionally
considered stakeholders, be included in discussions. Institutions that promote inclusiveness
can arrive at solutions that otherwise may seem unattainable.

Modern transboundary water agreements typically exhibit concepts of fairness
and of ecological sustainability taking precedence over economic efficiency (Syme
et al. 1999; Wolf 2000; Blomquist and Ingram 2003; Brandes and Curran 2016).
Careful review of even some of the more contentious river basins shows that the
extent of cooperation increases roughly in proportion to advances in other diplomatic
areas. In her review article, Selina Ho cites several articles on international water
issues as showing that "managing transboundary river basins is an exercise in foreign
policy-making and diplomacy that goes beyond the technical details of river basin
management" (2018, p. 621).

Of course, forming transboundary water agreements is never easy. Once one
merges water management with foreign policy-making, the course often veers toward
a large water hole that has been dubbed "hydrocentricity" (Brichieri-Colombi 2004).
The term connotes excessive emphasis on large-scale water infrastructure, such as
high dams, and on policy that places water at the centre of state building. Sensational
reporting sometimes presents fresh water as the key to Israeli national security or
Palestinian economic development, but those statements are exaggerations. None of
the recent books on the Israeli–Palestinian conflict gives more than minor attention
to the role of fresh water as a decisive issue between the two sides though most
do say that the Palestinian community needs more water than it currently gets, a
point that is not seriously in question. Fresh water is simply not a major issue in the
Israeli-Palestinian conflict.

In conclusion, Grover and Krantzberg (2015) emphasize that the International
Joint Commission (IJC) was originally formed to ensure just and workable allocations of water bodies along the Canada-US border. However, over time it has come to
recognize the importance of an ecosystem approach. It is well past time for Israelis,
Jordanians and Palestinians to begin thinking of their shared waters from a hydrosocial or political ecological perspective (Boelens et al. 2016).

7.2 General Conclusions

EcoPeace has long maintained that water issues *need not* wait. We now assert that
they *cannot wait* and they *should not* wait. They cannot wait because under the
existing situation neither side is making the best use of its fresh water, with adverse
results that range from economically costly to ecologically destructive. They should
not wait because an agreement to share water peacefully will be a model to show that
agreements on other issues can be reached between Israelis and Palestinians. Though
looking toward a Final Status Agreement, the EcoPeace Proposal is designed in a
way that allows it to be adopted prior to that Agreement. Only minor adjustments
would be required when final borders are established.

However, to now the systematic focus of both Israeli and Palestinian governments on water quantities has created bottlenecks in water negotiations (One can argue that it is only thanks to the geological rift that Israel and Jordan escaped a similar bottleneck). Each party sought to secure a stock of water, which then structured its approach to negotiations and dictated the categories deployed to describe reality. Both parties ignored that the water quantities they discuss flow through several users' hands, both Israeli and Palestinian, between source and sink, and that the same unit of water may serve different uses.

With its massive desalination capacity, Israel now enjoys a modest water surplus. This is an ideal situation from which to reconsider its interaction with its Palestinian neighbour as it already has with Jordan. Palestinians are located upstream on the aquifers; Israelis are located upstream for desalination capacity; Jordanians are located upstream for geography. Reformulating their negotiations from a struggle over quantities to a one over interactions with a flow is now not just possible but desirable.

Our final conclusions come at two levels. First, to the extent that the Israel-Palestinian differences depend on conflicts over water, we argue that when Israel passed its 1959 Law on Water, many people maintained that it had created the world's first modern water law (Tal 2002; Trottier 1999). If Israeli, Palestinian, Jordanian negotiators adopt the EcoPeace Proposal for joint management of shared water, we believe they will have created the world's first *post-modern* water agreement, or, as we prefer to term it, a new paradigm for managing transboundary water.

Turning now to the global level of concern, though water security is critical to any society, securitizing water (in the sense of asserting a direct link between available water and either military strength or standard of living only gets in the way of efficient, equitable, sustainable, and implementable development of shared water resources. To the contrary, water security in the form of institutions that provide both peoples with adequate water for comfortable lifestyles and productive livelihoods will have to be efficient, equitable, sustainable, and implementable if they are not to fail one or the other side, or more likely both sides.

If Israelis and Palestinians can come to agreement over the water they share, there is no reason why a similar process might not apply elsewhere in the Middle East, and indeed elsewhere in the world where transboundary water divides rather than unites communities or states on opposite sides of a border.

References

Bazza M, Kay M, Knutson C (2018) Drought characteristics and management in North Africa and the Near East. FAO, FAO Water Reports 45, Rome

Blomquist W, Ingram HM (2003) Boundaries seen and unseen: resolving trans-boundary ground-water problems. Water Int 28(2):162–169

Boelens R, Hoogesteger J, Swingedouw E, Vos J, Wester P (2016) Hydrosocial territories: a political ecological perspective. Water International 41(1):1–14

Brandes OM, Curran D (2016) Changing currents: a case study in the evolution of water law in western Canada. In: Renzetti S, Dupont DP (eds) water policy and governance in Canada. Springer, New York, pp 45–67

Brichieri-Colombi JS (2004) Hydrocentricity: a limited approach to achieving food and water security. Water Int 29(3):318–328

Gafny S, Talozi S, Al Sheikh B, Ya'ari E (2010) Towards a living Jordan River: an environmental flows report on the rehabilitation of the Lower Jordan River. Friends of the Earth Middle East, Amman, Bethlehem, Tel Aviv. http://foeme.org/uploads/publications_publ117_1.pdf

Grover VI, Krantzberg G (2015) Transboundary water management: Lessons from North America. Water Int 48(1):183–198

Hadi MA (ed) (2003) Water in Palestine: Problems, politics, prospects. PASSIA Publications, Jerusalem

Hellegers P, Leflaive X (2015) Water allocation reform: what makes it so difficult? Water Int 40(2):273–285

Ho S (2018) Introduction to special section: transboundary river cooperation: Actors, strategies and impact. Water Int 43(5):620–621

Megdal SB, Varady RG, Eden S (eds) (2013) Shared borders, shared waters: Israeli-Palestinian and Colorado River Basin water challenges. CRC Press/Balkema, Leiden, Netherlands

Sagive M et al (2012) Community-based problem solving on water issues: cross-border "priority initiatives" of the Good Water Neighbors Project. Friends of the Earth Middle East, Amman, Bethlehem, and Tel Aviv

Shuval HI (2007) Meeting vital human needs: Equitable resolution of conflicts over shared water resources of Israelis and Palestinians. In: Shuval HI, Dweik H (eds) Water resources in the Middle East: Israeli-Palestinian water issues—from conflict to cooperation. Springer, Berlin, pp 1–16

Shuval HI, Dweik H (eds) (2007) Water resources in the Middle East: Israel-Palestinian water issues—from conflict to cooperation. Springer, Berlin

Syme GJ, Nancarrow BE, McCreddin JA (1999) Defining the components of fairness in the allocation of water to environmental and human uses. J Environ Manage 57(1):51–70

Tal A (2002) Pollution in a promised land: an environmental history of Israel. Univ of California Press, Berkeley, CA, USA

Trottier J (1999) Hydropolitics in the West Bank and Gaza Strip. PASSIA Publications, Jerusalem

Wildeman J Tartir A (2019) Why cutting us aid will help Palestinians—and peace. Middle East Eye, 06 February 2019

Wolf AT (1999a) 'Water wars' and water reality: conflict and cooperation along international waterways. In: Lonergan S (ed) Environmental change, adaptation, and security (NATO ASI series vol 65). Kluwer Academic Press, Dordrecht, Netherlands

Wolf AT (1999b) Criteria for equitable allocations: The heart of international water conflict. Nat Resour Forum 23(1):3–30

Wolf AT (2000) From rights to needs: water allocations in international treaties. In: Feitelson E, Haddad M (eds) Management of shared groundwater resources: the Israeli-Palestinian Case with an international perspective. International Development Research Centre, Ottawa, and Kluwer Academic, Amsterdam, pp 27–59

Afterword

This book focuses on water issues, water management, and water governance in Israel, Palestine, and Jordan with emphasis on negotiations, technologies, and social and political changes since 2000. Obviously, there was a lot of water history before that time, and the two annexes deal with those earlier periods of time: Annex A from the initial years of the British mandate over what was then called Palestine through Israeli independence and a bit beyond; Annex B picks up where Annex A ends and continues through the active period of efforts to resolve international issues during the 1990s. A second section of Annex B reviews the role of the Water Resources Working Group, and a third section deals with the impacts of climate change on modern thinking about water.

D. B. Brooks et al., *Transboundary Water Issues in Israel,*
Palestine, and the Jordan River Basin, SpringerBriefs on Case Studies
of Sustainable Development, https://doi.org/10.1007/978-981-15-0252-1

Annex A: Water Studies During and Immediately After the British Mandate (1922–1950s) [1]

The management of reticulation (local piped water) networks in Jerusalem was already the object of international politics in the 19th century (Lemire 2011). However, until the 1920s the bulk of water used in the region went to irrigation and was the object of local politics only. The perception of water in the Jordan Basin as an international problem of quantitative allocations arose in the 1920s when the French and British governments established their mandates over the remains of the Ottoman Empire.

This annex will describe the series of attempts to reach agreements over water resource availability in Israel-Palestine or, in some cases, in the Jordan River basin. Discussion begins with the British-Palestine Mandate of 1922, which established the region (excluding Trans-Jordan) as a distinct political unit. It also marked official international recognition of the historical connection of the Jewish people with the land of Palestine, and it spawned the development of a Jewish agency to assist with the administration of Palestine. Box A.1 lists the sequence of activities and proposals related to water management over the period 1922–1955.

Alatout (2007) has documented the tendency for Zionist proposals prior to statehood to emphasize the abundance of water for new immigrants, whereas those written after statehood emphasize the scarcity of water. However, the purpose of this annex is not to analyze the different studies but simply to note those differences, which depend significantly on the perspectives of the authors and/or of the agencies for which they worked.

> **Box A.1: Major Proposals Related to Sharing Water in Israel-Palestine or in the Jordan Basin Prior to 1960**
>
> 1922: British-Palestine Mandate established (no mention of water)
>
> 1926: Rutenberg Concession granted for hydroelectric station just below confluence of Jordan and Yarmouk rivers
>
> 1936: First regional water project delivered water to western Galilee
>
> 1937: Mekorot founded as Israel's National Water Company
>
> 1939: Ionides Plan (favoured by the Arabs)
>
> 1944: Lowdermilk Plan published (favoured by the Zionists)
>
> 1948: Hays published *TVA on the Jordan*, Proposals for Irrigation and Hydro-Electric Development in Palestine
>
> 1951: Murdoch MacDonald Corporation Report commissioned by the Government of Jordan
>
> 1952: Bunger Report commissioned by the Government of Jordan and the US Technical Cooperation Agency

---

[1] Annex A is adapted from a similar annex in the revised version of the EcoPeace Proposal (Brooks and Trottier 2012).

1952: Main Plan (based on TVA model), called by some the Unified Plan
1953: Israel begins construction of National Water Carrier
1953: Initial Johnston Plan distributed to the riparian states
1954: Cotton Plan (Israeli response to Initial Johnston Plan)
1954: Arab League's Technical Committee Plan (Arab response to initial Johnston Plan)
1955: Modified Johnston Report (sometimes also called the Unified Plan) published

A.1 1920s

Though it was not part of the mandate, in the minds of many Zionists the British-Palestine mandate document implicitly provided for an independent Jewish state (Lonergan and Brooks 1994). Soon after, a number of national development agencies and projects were created, including the Jewish-owned Palestine Electricity Corporation, which was founded by Pinhas Rutenberg. In 1926, British authorities granted the corporation a 70-year concession to the waters of the Jordan and Yarmouk rivers for the purpose of generating electricity, and subsequently a dam was built at the confluence of the two rivers. It was through this concession that Arab farmers were denied the right to use the waters upstream of the junction of the two rivers for any purpose without the permission of the Electricity Corporation, permission that was never granted (Isaac and Hosh 1992). Although the hydroelectric plant was damaged and ceased to operate following the 1948 war, Wolf (1995) says that Israel later used the Rutenberg concession to argue for a greater share of Yarmouk River water.

A.2 The 1930s

During the first half of the 1930s, the specific issue of fresh water availability became secondary to more general questions about the capacity of the land to support a higher population. Not surprisingly, concerns about the absorptive capacity of Palestine grew as Jewish immigration and settlement in the region and, concurrently, Arab opposition, increased. The first regional water supply project in Palestine was implemented in 1935–36 and involved supplying water to the western Galilee. After this project, the British assigned Michael George Ionides to be Director of Development for the East Jordan Government for the express purpose of assessing the water resource and irrigation potentials of the Jordan River Basin. The Ionides Plan contained three primary recommendations:

- Yarmouk River floodwaters would be diverted along the East Bank of the Jordan River and stored in the Sea of Galilee
- Stored water, along with a small quantity of Yarmouk River water, would be diverted to a new canal (the East Ghor Canal) to provide irrigation for lands east of the Jordan River
- Irrigation water of the Jordan River would be used primarily within the Jordan River Basin.

In 1938, Walter Clay Lowdermilk, a director of the US Soil Conservation Service, was sent to the region to examine the potential for greater land conservation. He felt that, with appropriate management, the water available in the Jordan River Basin could sustain a much larger population than existed at that time. His initial idea included the formation of a regional water authority based on the Tennessee Valley Authority (TVA), which at that time was considered a great success in the United States and thus appropriate for other locations (Lonergan and Brooks 1994).

A.3 The 1940s

In 1944, Lowdermilk published his comprehensive plan for the region, entitled *Palestine: Land of Promise*. The plan proposed that, by exploiting unused water resources adjacent to Palestine, particularly the Litani River (in Lebanon) and the Yarmouk River, water could be diverted for irrigation throughout the Jordan Valley and south to the Negev. However, there was a major problem with use of the TVA as a model: it ignored the social capital produced by local property rights systems that were used to manage irrigation water in most of the region.

A few years after its founding in 1937 as the water agency for Jewish villages and cities in Palestine, Mekorot also prepared a plan for resolving the water resource problems. Its plan proposed a "national" water resource project that focused on irrigation and hydroelectric development, and incorporated both surface water (from the Yarmouk, the Yarkon, and the Jordan, as well as springs and floodwaters) and ground water. The plan had an element of expansionism in that it also suggested that the Mandate border be redrawn to include the three headwaters of the Jordan River: the Hasbani River (mainly in Lebanon) and the Banyas Stream (from the foot of Mount Hermon/Jabel Sheich on the flanks of the Golan Heights), as well as the Dan Stream, which was already within the Mandate. As well, the plan suggested that the Mandate border be extended eastward to include territory for a conduit along the shores of Lake Hula and upstream on the Yarmouk River (affecting both Syria and Jordan) to allow for a set of impoundments to store water for irrigation (Wolf 1995).

Zionists strongly supported both the Lowdermilk and Mekorot plans. The World Zionist Organization asked James B. Hays, an engineer who had worked on the TVA in the United States, to draw up development plans based on Lowdermilk's ideas. Hays agreed with Lowdermilk's arguments about the capacity of Palestine to support a larger population, and he published his plan in a book entitled *TVA on the Jordan*. This plan comprised seven elements:

- Development of groundwater resources
- Development of the Upper Jordan River's summer flow for irrigation of nearby lands (including diversion of the Hasbani River for irrigation)
- Diversion of Yarmouk River waters into the Sea of Galilee and their storage there
- The Mediterranean Sea–Dead Sea ("Med-Dead") Canal that had first been proposed by S. Blass, then working with Mekorot
- Recovery of the Jordan River's winter flow for irrigation of the coastal plain
- Reclamation of the Hula marshes (an area flooded by winter flow from the Jordan River) by constructing a series of drainage canals to control flood water, recharge aquifers, and convert the marsh into fertile irrigation land

- The use of flood water for irrigation in the Negev.

The disagreement as to the number of people the region could support and the types of water projects needed to provide for population growth was never resolved. Instead, the United Nations Partition Plan of 1947 and the subsequent 1948 War changed the locus for decision-making and set the stage for water conflicts over the next few decades.

A.4 The 1950s

The British Mandate for Palestine of course ended in 1948 when a United Nations vote divided the territory into Palestinian and Jewish areas. Immediately thereafter the State of Israel was created in the latter, and along with it Yom haAtzmaut (Independence Day) for Jews around the world, and Nakba (catastrophe or disaster day) for Palestinians around the world to commemorate the expulsion or flight of more than 700,000 Palestinians during the subsequent war.

To no one's surprise, the development of water resources continued to play a major role in national policy making in the new State of Israel when immigration was very high and job creation not high enough to absorb them all. The one sector that could use many of the new Israelis was agriculture, and in the first decade after independence (1948–1958) the land cultivated by Jews rose from 160,000 to 390,000 ha (de Chatel 2007). In roughly the same period, agricultural output increased by a factor of five (*Ibid.*), something that could not have been possible in the absence of irrigation.

The first formal plan for water management in the post-independence period in Israel was the MacDonald Report in 1951 (Wishart 1990). This report outlined the conflicts between Jordan and Israel and proposed that any water withdrawn from sources in the Jordan Valley remain in the valley. The proposal also included the Hays Plan component of diverting the Yarmouk River into the Sea of Galilee (Isaac and Hosh 1992). However, the Arab states were concerned about sharing a reservoir with Israel, even though it was a much cheaper alternative than building independent storage capacity outside Israel (Kally with Fishelson 1993).They favoured a plan proposed by M. Bunger, an American engineer working in Amman for the Government of Jordan, which involved the construction of a high dam on the Yarmouk to provide water storage and hydroelectric capacity. The dam was to be built at Maqarin as a joint project between Jordan and Syria. It would also use the winter flow from the Yarmouk to generate electricity for both Syria and Jordan, with 75% going to Syria (Wishart 1990). Construction of the dam began in 1953, but Israel raised strong objections to development of the Yarmouk because it would affect flows into the Jordan River, and pressured the United States to withdraw funding for the plan (Isaac and Hosh 1992).

Anticipating that bringing several of these proposals together might alleviate some of the conflicts among riparians on the Jordan, the United Nations Relief and Works Agency asked the TVA to develop a "unified plan." In 1952, the TVA requested Charles T. Main, Inc. to combine all the work previously conducted by the various parties into one plan. Borrowing the key objectives of the earlier Ionides and Mac-Donald proposals, the Unified Plan was based on irrigation by gravity flow, which implies that all water will be used within the watershed where it originates. It also included drainage of the Hula marshes, storage of Yarmouk River water in the Sea of Galilee, a Med-Dead Canal proposal, and dams on the Hasbani Stream and Yarmouk River for irrigation and power.

In parallel with discussions about a regional water plan, Israel undertook some unilateral projects in the Jordan River basin. In 1953, Israel began construction of its National Water Carrier (see Fig. 2.2) at a site in the demilitarized zone north of the Sea of Galilee. Syria responded by sending troops to the border and, according to Cooley (1984), firing artillery shells at the construction site. Syria also protested to the United Nations, and the Security Council responded by ordering that work in the demilitarized zone be halted. Israel then moved the intake site for the National Water Carrier to the Sea of Galilee, a move that, as Wolf (1995) notes, was "doubly costly" for Israel. The salinity of the Sea of Galilee was higher than that of the Upper Jordan; as a result Israel had to divert saline springs away from the lake and into the Lower Jordan. In addition, the water now had to be pumped up 250 m from the intake location before heading southward and eastward.

Although tensions had been temporarily relaxed by the Israeli decision to move the intake site for the National Water Carrier, the pressing need for a regional solution to problems involving the Jordan River remained. As well, pressure was increasing to resolve the issue of Palestinian refugees. As a result, in 1953 Eric Johnston was appointed by US President Eisenhower as a special ambassador to lead a mission to propose multilateral water development of the Jordan River basin. Shuttle diplomacy among Syria, Lebanon, Jordan and Israel took place over the next two years. The West Bank was included in Jordan's share of water, and the Gaza Strip was ignored because it was then part of Egypt and not in the Jordan Valley. In this case, diplomacy came very close to succeeding, as described in a pair of articles by Phillips and colleagues (2007a, b). What came to be known as the Initial Johnston Plan had three major components:

- Water storage including proposals from earlier studies to construct a dam near Maqarin and a diversion structure to store winter flows from the Yarmouk River in the Sea of Galilee.
- Water distribution focused primarily on providing water to Jordan's East Ghor Canal, which would then supply most of the surface water to that country.
- Water allocations were based on ensuring that Arab states receive enough water to meet their irrigation needs, with the remaining water divided between Jordan (the Yarmouk) and Israel (the Jordan).

Not surprisingly, the Initial Johnston Plan was not acceptable to either Israel or to the Arab states. Israel considered the allocations it was to receive under the plan insufficient and argued that a regional plan should include all water sources of the region, including the Litani River in Lebanon. The Arab states remained concerned about the storage of Yarmouk River water in the Sea of Galilee as well as the high allocation given to Israel. Accordingly, both groups prepared alternative proposals. The Israeli proposal, prepared by Joseph Cotton, an American engineer, included an allocation to Israel of 55% of Litani and Jordan waters (compared with 33% under the Main Plan). The Cotton Plan also allowed for the use of Jordan River water outside the watershed (for irrigation in the Negev). The Arab League's Technical Committee Plan was consistent with the Main Plan in that it required that all water be used within its watershed, but it reduced Israel's share to 20% and did not include the Litani River. In spite of these differences, all of the parties recognized the need for regional cooperation for efficient utilization of water resources. The primary disagreements were limited to the size of water allocations and the transfer of water outside the watershed (Lonergan and Brooks 1994).

Using the two counterproposals, along with a recently completed hydro-graphic survey commissioned by the Jordanian government, Eric Johnston submitted a revised set of proposals in 1955. This Modified Johnston Plan allowed for inter-basin transfer within the context of the allocations to each country and incorporated many of the engineering features of the Main Plan. However, disagreements remained over allocations and international supervision. The Arabs were in favour of direct supervision by an international body, whereas Israel preferred supervision by a group of engineers from the region. Even so, by late1955 Johnston could report that: "They [the riparian states] have made it clear ... that the technical and engineering aspects of the plan ... are now satisfactory to them," and went on to say that the negotiations had reached the "one inch line" (as cited in Garbell 1965). Israel did grant formal political support to the Modified Johnston Plan, and it was accepted by the Arab League's Technical Committee (Haddadin 2011). However, the plan was never formally implemented, largely because Arab states feared that their signature might be taken to imply formal recognition of Israel, which at the time was unacceptable to them (Lonergan and Brooks 1994). Wishart (1990) concludes that the Arab states had little to lose by not entering into the agreement. In practice, all of the riparian states unofficially accepted the Modified Johnston Plan, with the exception of Syria, which did not reject it, but simply failed to act on it.

Looking at the same data after some 50 years, Phillips and his colleagues (2007a, b) agree that the 1955 Modified Johnston Plan does make sense in terms of more recent international water law. However, they caution that its workability depended upon creating additional sources of water through desalination, wastewater recovery, or imports. It also recommended that water banking be introduced as a way to share the gain in years of adequate rain and the pain in years when rainfall is less than adequate.

For a time the Modified Johnston plan provided a workable arrangement for water sharing for the Jordan Basin, but, in the absence of formal adoption, its influence gradually declined (Lonergan and Brooks 1994; Elmusa 1996). Today, it is doubtful whether either Israel or Syria would accept the Johnston Plan because each has built its water system in ways that give them more water than their allotments under that proposal. Any Palestinian government would certainly reject any version of the Johnston plan because, with most Palestinians living in areas then under Jordanian control, the Palestinian share was simply included in the Jordanian share. The Eco-Peace Proposal also rejects the Johnston Plan for three reasons: first, as indicated in Sect. 4.2 above, the approach of fixed quantitative allocations is misguided; second,

It lacks a clear Palestinian share; and, third, all water in the Jordan Basin is treated as if available for human uses with none left for ecosystems.

By way of conclusion, it is also noteworthy that Elisha Kally with Gideon Fishelson (1993), both prominent Israeli engineers, proposed a number of more or less visionary schemes to bring fresh water to Israel from the Nile (Egypt), the Yarmouk (mainly in Syria), or the Litani (Lebanon). The projects are shown to be technically feasible but, with one or two exceptions involving the Jordan River, rather expensive, even if all the parties are supportive, which of course they were not. The book is a useful compendium of what might be possible, but it fails as a guide to policy. Based on a review of the text, it does not appear that the words "conservation," "efficiency," or "environment" appear anywhere.

Annex B: Water, Wars, and the Israeli-Palesinian Conflict (1950s to 2000)

In the Epilogue of his careful review of "Water Disputes in the Jordan Basin Region," Libiszewski (1995, pp. 91–95) focuses on "Weighing Water's Role in the Arab-Israeli Conflict." He adopts Dressler's analytical approach (1994) of "distinguishing four different roles that 'causes' may play in generating and sustaining violent conflict":

- As Triggers: actions that increase the probability of violence.
- As Targets: a decision-maker's objective, aim, or goal.
- As Channels: lines of political, social, economic, or national cleavage among groups.
- As Catalysts: any factor that controls the rate or intensity and the duration of a conflict.

Libiszewski then tests the numerous conflicts, including the 1967 war between Israel and its Arab neighbours, to assess what causal role, if any, that water played. With minor exceptions, he is looking for proximate causes, not distant or otherwise limited ones. He also has a second goal, that of disputing the so-called "hydraulic imperative" that is alleged to have propelled the Israeli government to react positively to those causes.

B.1 Dressler-Libiszewski Analysis

There is no reason to think that water was a trigger for Arab-Israeli conflict. The trigger rather was the effort by Zionists to build a Jewish state on land that may have been historically Jewish but that for nearly two millennia had been settled by Palestinian Arabs. Though water resources were certainly high among issues for Zionist planners (see Annex A), any subsequent conflicts "were rather an outflow of political and territorial conflict rather than part of its origin" (p. 92). Once the State of Israel came into existence, "water became a critical factor of economic development for all parties involved," and, because most of the larger water resources are transboundary, "competition of shared water turned into one of the proximate triggers of conflict and violence." Notwithstanding that general potential, it did not apply to all conflict. Notably, "Outbreak of the Six Days' War itself was not directly triggered by events related to water," and often enough water was just "a means of carrying out the deeper historical conflict."

In contrast to its ambiguous role as a trigger, water infrastructure has certainly been a political or military target, particularly during the first 25 year of Israel's existence, most often as a secondary target in a broader struggle over territory. Both sides shot at dams or construction sites for water, as when in 1951 the Syrian army threatened Israeli plans to drain the Hula swamp in the north of the country to provide a source for the National Water Carrier, or when in 1965 and 1967 the Israeli army and air force attacked work sites for an Arab plan to divert the Hasbani and the Banias (the two non-Israeli tributaries to the Jordan River). However, these skirmishes are best seen as a form of diplomacy designed to indicate that the intended construction would affect vital resources and therefore induce the other side to "back off." Together with international pressure, the inducement was more often than not successful. (As noted in Annex A, moving the intake for the National Water Carrier from the Hula swamp to the northern part of the Sea of Galilee was costly for Israel as the Hula swamp is about 70 m above sea level whereas the Sea of Galilee is 250 m below sea level, which meant adding 320 extra metres of pumping for the National Water Carrier.)

In summary, for about 20 years, hydraulic installations were among the preferred targets "for actions aimed at weakening or castigating the enemy," but "this link must be regarded as a military *instrument* rather than as a causing dimension of the conflict" (p. 94), a point that Libiszewski makes more strongly in an earlier paragraph (p. 93).

> Some authors have maintained the thesis that, driven by a 'hydraulic imperative,' capture of additional water resources was a primary motive for Israel to go to war in 1967 and 1982 (*e.g.* Cooley 1984). This is surely a too simplistic interpretation of the matter, . . . /Yet/, 'although water may not have been the prime impetus behind the Israeli acquisition of territory, as the hydraulic imperative alleges, it seems to be perhaps the main factor determining its retention of that territory" (Frey and Naff 1985, p. 76).

In addition to this cautious evaluation of any link between water and the 1967 Six Day War, there exists a well-argued case some years later for some channelling from drought conditions to violence. For example, in Syria's Euphrates River basin the displacement of many farmers for lack of irrigation water is commonly linked to the country's subsequent civil war and invasion by Islamic extremists.[2] Any direct cause-and-effect is no doubt also simplistic, but that does not eliminate the argument for channelling, as suggested in numerous articles and websites, most carefully in Gleick (2014) and in an article in *Smithsonian Magazine*.[3]

B.2 The Water Resources Working Group (WRWG) of the Middle East Peace process[4]

In October 1991, a three-day conference aimed at finding some resolution to the conflicts between Israel and its Arab neighbouring states was held in Madrid under the sponsorship of Spain, the United States, and Russia (at the time, the USSR). From the start, it was recognized that the participants did not have the power to impose or veto a solution, but they all had an interest in getting away from ongoing conflict in the region. At a follow-up conference in January 1992 in Moscow it was agreed to create two groups of activities: the Bilateral Track and the Multilateral Track. The Bilateral

---

[2] See Chap. 1 in de Chatel (2007) for a description of conditions in the years just before the civil war.

[3] https://www.smithsonianmag.com/innovation/is-a-lack-of-water-to-blame-for-the-conflict-in-syria-72513729/

[4] Comments in this annex are based in part on a review of the literature and in part on the fact that David Brooks, employed at the time by Canada's International Development Research Centre, was often called upon to accompany diplomats from the Canadian Department of Foreign Affairs and International Trade to the meetings of the WRWG (as well of the Environment Working Group) as an "expert." He was therefore present at many of the formal and some of the intersessional meetings that took place between 1992 and 2000.

Track involved separate talks between Israel and each Jordan, Lebanon, Syria, and the Palestinians with oversight from the United States and Russia and with emphasis on political issues. The Multilateral Track included five working groups that, it was hoped, would get Israelis, Jordanians, and Palestinians from the West Bank and Gaza Strip used to joint efforts on five substantive topics: water resources, environment, arms control, refugees, and economic development. Emphasis was to be placed on technical issues and, in contrast to the Bilateral Track, participation by a large number of regional and non-regional states was encouraged. Syria and Lebanon refused to take part in the Multilateral Track on the basis that political issues had to be resolved before technical ones could be considered. Few other regional states took this same position.

The remainder of this annex will focus on the Water Resources Working Group (WRWG) for which the United States was gavel holder. (There is a vague but apparently important distinction between the terms "gavel holder" and "chairperson" in diplomatic discussions.)[5] Early on, the United States together with Japan and the European Union as co-organizers proposed four agenda items to guide future meetings and activities of the WRWG:

- Enhancement of water data availability
- Water management practices, including conservation
- Enhancement of water supply
- Concepts of regional water management and cooperation.

The conclusion from the Transboundary Freshwater Dispute Database at Oregon State University, which provides a meeting-by-meeting review of events and results from the WRWG, is quite positive:

> Given the length of time that the region has been enmeshed in bitter conflict, the pace of accomplishment of the peace process has been impressive, no less so in the area of water resources. This may be due in part to the structure of the peace talks, with the two complementary and mutually reinforcing tracks—the bilateral and the multilateral. As noted earlier, past attempts at resolving water issues separate from their political framework, dating from the early 1950s through 1991, have all failed to one degree or another. Once the taboo of Israelis and Arabs meeting openly in face-to-face talks was broken in Madrid in October 1991, the floodgates were open, as it were, and a flurry of long-repressed activity on water resources began to take place outside the official peace process.

There is no reason to doubt this assessment, but it is also true that throughout its life the WRWG was bedeviled by the lack of definition of what was a technical issue focusing on joint management and new resources, as the Israelis preferred to think, and what was also a political issue focusing on water rights and existing grievances, as the Palestinians preferred to think. This problem was never adequately resolved,

---

[5]Much of the material on the WRWG is based on a website from the Israel's Ministry of Foreign Affairs (https://mfa.gov.il/mfa/foreignpolicy/peace/regional/pages/middle%20east%20multilateral%20working%20group%20on%20water%20re.aspx), or from the Transboundary Freshwater Dispute Database at Oregon State University. (https://transboundarywaters.science.oregonstate.edu/sites/transboundarywaters.science.oregonstate.edu/files/Database/ResearchProjects/casestudies/middle_east.pdf).

but neither did it prevent some important gains being made in water resources management for the region, particularly after 1993 when agreement was reached that the WRWG would emphasize plans for the future rather than implementation in the present. The Working Group did not stick strictly to this distinction, but the agreement was enough to allow work to proceed on each of its four guiding agenda items.

Notable among the gains in the first agenda item on enhancement of water data availability was a project that was known by the name of EXACT, which was short for Executive Action Team, and was composed of two members from each regional party and two representatives from each Donor Party. Its prime objectives were not just to improve water data collection and management, but also to bring these data, many of which were state secrets, into a common framework so they could function effectively in a regional setting and also meet specific needs of each party. EXACT has been far more successful than expected at creating a regional water data bank that is now regularly used by the Palestinian Water Authority, the Israeli Hydrological Service, and the Jordanian Ministry of Water and Irrigation.

A second major success was the establishment of the Middle East Desalination Research Center (MEDRC) in Muscat, Sultanate of Oman, in 1996, in response to the agenda item on Enhancement of Water Supply. Its mission was (and still is) to conduct, facilitate, promote, coordinate, and support basic and applied research in water desalination and supporting fields in an effort to reduce costs and improve quality in the technical processes of water desalination for both sea water and brackish water.

Dozens of other projects were implemented over the years in response to the other two agenda item. Most of them were study tours or training courses. Others were studies of water supply-demand gaps in specific regions undertaken by one of the participating states. Perhaps, the most useful was a German Government study of alternative scenarios for water resources in the short, medium, and long terms. To no one's surprise, the study showed a significant gap between water supply and demand throughout the region, even when using conservative estimates of future population growth and water use. However, the study was able to identify the main sources of the gap and the most likely options for reducing it. Other studies compared specific water management policies, including water pricing, among countries across the Middle East.

Only a few projects were actually designed to deal with existing supply-demand problems, and they were all small in scale, and thus implementable without stepping on political toes. A good example was the Canadian project to show how rooftop rainwater catchment systems in the Gaza Strip could help alleviate its chronic shortage of potable water; it actually built about eight demonstration systems.

Despite being named as one of the four agenda items, projects on water conservation were notable mainly by their absence. Some work on public education was done, a teacher' guide was produced, and an experimental farm started, but no project actually put water conservation activities into practice. This absence leads one to suspect that, for whatever reason, reducing water demand was seen as political whereas increasing water supply was not.

Most of the activities of the WRWG were terminated around the end of the millennium. The call by the Arab League for a boycott of the multilaterals as a protest against Israeli policies had halted some projects, and the Second Intifada in September 2000 ended most of the rest. As with so many efforts before, the Multilateral Track too was vulnerable to disruptions in the broader Middle East peace process. However, specific projects, such as MEDRC, and new institutions that could find their own funding were able to survive, and they continue to contribute to improved water management in the region.

B.3 Climate Change Begins to Play a Role (1991–2000)

Libiszewski's analysis ends in 1995 when his monograph was published. Population was not yet a major problem and climate change was just beginning to become of concern. Yet, even then he warned that as populations continued to grow, and as climate change began to affect adversely water resources in the countries of the Jordan River basin, water might turn out to play a greater role and then become a serious cause of conflict.

Recently, *The Gallon Environment Letter*, a Canadian environment and business periodical gave some startling dimensions to the effect humans have had on our ecology over just the past 50 years:[6]

land, marine and aquatic areas are being destroyed by humans: natural forests lost at the rate of 6.5 million hectares each year from 2010 to 2015; natural wetlands declined by 35% from 1970 to 2015; 30% of corals are at risk of bleaching which could lead to their death and that of life depending on corals, 60% of vertebrates have disappeared since 1970.

Petersen-Perlman et al. (2017) and Inga (2019) are two of the recent analyses that focus on potential future changes in water availability as a result of climate change. Both conclude that all three Jordan Valley countries will find it difficult to meet existing water commitments, much less larger ones to be expected with immigration and refugees in the future. More to the point, none of Israel, Palestine, or Jordan is building climate change strongly enough into its planning nor recognizing the need for adaptive capacity to preserve regional cooperation.

In sum, Israel, Palestine, and the western part of Jordan are going to have to plan carefully and use water modestly to survive in a future dominated by climate change. Fortunately, they all have the human abilities and the economic capital to succeed. But, equally or perhaps even more important, do they have the political will to overcome past conflicts and work together for a better shared future?

---

[6]Vol. 21, No. 6, July 31, 2019; https://mail.google.com/mail/u/0/?tab=wm#inbox/FMfcgxwDqnjnHRBjlZTbgxzDVxNQBXGc

References

Alatout ST (2007) From water abundance to water scarcity (1936–1959): a "fluid" history of Jewish subjectivity in historic Palestine and Israel. In: Sandy Sufian S, LeVine M (eds) Reapproaching borders: new perspectives on the study of Israel-Palestine. Bowman and Littlefield, Lanham, MD, USA, pp 199–219

Cooley JK (1984) The war over water. Foreign Policy 54(Spring):3–26

De Chatel F (2007) Water sheikhs and dam builders: stories and people and water in the Middle East. Transaction Publishers, New Brunswick, New Jersey, USA

Dressler D (1994) How to sort causes in the study of environmental change and violent conflict. In: Graeger N, Smith D (eds) Environment, poverty, conflict. PRIO Report No. 2, pp 91–112

Elmusa SS (1996) Negotiating water: Israel and the Palestinians. Institute for Palestinian Studies, Beirut

Frey FW, Naff T (1985) Water: an emerging issue in the Middle East. In: The annals of the American academy of political and social science 482:65–84

Garbell MA (1965) The Jordan valley plan. Sci Am 212(3):23–31

Gleick PH (2014) Water, drought, climate change, and conflict in Syria. Water, Climate and Society 6:331–340

Haddadin MJ (2011) Water: triggering cooperation between former enemies. Water Int 36(2): 178–185

Inga C (ed) (2019) Climate change, water security, and national security for Jordan, Palestine, and Israel. EcoPeace Middle East, Amman, Ramallah, Tel Aviv

Isaac J, Hosh L (1992) Roots of the water conflict in the Middle East. Applied Research Institute, Jerusalem

Kally E, with Fishelson G (1993) Water and peace: water resources and the Arab-Israeli peace process. Praeger. Westport, CT, USA

Lemire V (2011) La soif de Jerusalem: Essai d'hydrohistoire (1840–1948). Publications de la Sorbonne, Paris

Libiszewski S (1995) Water disputes in the Jordan basin region and their role in the resolution of the Arab-Israeli Conflict. Center for Security Studies and Conflict Research, Swiss Federal Institute of Technology, Zurich, and Swiss Peace Foundation, Bern

Lonergan SC, Brooks DB (1994) Watershed: the role of fresh water in the Israeli-Palestinian conflict. International Development Research Centre, Ottawa

Petersen-Perlman JD, Veilleux JC, Wolf AT (2017) International water conflict and cooperation: challenges and opportunities. Water Int 42(2):105–120

Phillips DJH, Attili S, McCaffrey S, Murray JS (2007a) The Jordan River basin: 1. Clarification of the allocations in the Johnston Plan. Water Int 32(1):16–38

Phillips DJH., Attili S, McCaffrey S, Murray JS (2007b) The Jordan River basin: 2. Potential future allocations to the co-riparians. Water Int 32(5):39–62

Wishart D (1990) The breakdown of the Johnston negotiations over the Jordan waters. Middle East Studies 26(4):536–546

Wolf AT (1995) Hydropolitics along the Jordan River: Scarce water and its impact on the Arab-Israeli conflict. United Nations University Press, Tokyo

Bibliography

Eran O, Bromberg G, Milner M (2014) The water, sanitation and energy crises in Gaza: Humanitarian, environmental and geopolitical implications, with recommendations for immediate measures. Institute for National Security Studies, Tel Aviv, and EcoPeace Middle East, Amman, Ramallah, Tel Aviv

Fischhendler I (2015) The securitization of water discourse: theoretical foundations, research gaps and objectives of the special issue. Int Environ AgreemS: Polit, Law Econ 15(3):245–255

Fischhendler I, Dinar S, Katz D (2011) The politics of unilateral environmentalism: cooperation and conflict over water management along the Israeli-Palestinian border. Glob Environ Polit 11(1):36–61

Feitelson E, Tamimi A, Rosenthal G (2012) Climate change and security in the Israeli-Palestinian context. J of Peace Res 49(1):241–257

Gilmont M, Rayner S, Harper E et al (2017) Decoupling national water needs for national water supplies: Insights and potential for countries in the Jordan basin. R Sci Soc, Amman, WANA Institute

Gleick PH (2004) Environment and security: water conflict chronology, Version 2004–2005. In: Gleick PH (ed) The world's water 2004–2005: the biennial report on freshwater resources. Island Press, Washington, DC, pp 234–255

International Water Resources Association (2018) Sustainable groundwater development for improved livelihoods in Sub-Saharan Africa, Water International Policy Brief no. 9. Available at https://www.iwra.org/wp-content/uploads/2018/05/PB-N9-web-1.pdf

Le More A (2008) International assistance to the Palestinians after Oslo : political guilt, wasted money. Routledge, London

World Bank (2018) Towards water security for Palestinians: West Bank and Gaza. Water supply, sanitation, and hygiene poverty diagnostic. WASH Poverty Diagnostic, Washington, DC

Wouters P (2013) International law—facilitating transboundary water cooperation. TEC Background Papers No. 17, Stockholm: Global Water Partnership

Printed in the United States
By Bookmasters